AF360881

Wastewater Treatment, Valorization and Reuse

Wastewater Treatment, Valorization and Reuse

Editors

Mejdi Jeguirim
Salah Jellali

MDPI • Basel • Beijing • Wuhan • Barcelona • Belgrade • Manchester • Tokyo • Cluj • Tianjin

Editors
Mejdi Jeguirim
Université de Strasbourg, Université de
Haute-Alsace, CNRS, Institut de Science des
Matériaux de Mulhouse (IS2M) UMR 7361
France

Salah Jellali
Water Research and
Technologies Centre (CERTE),
Touristic Road of Soliman
Tunisia

Editorial Office
MDPI
St. Alban-Anlage 66
4052 Basel, Switzerland

This is a reprint of articles from the Special Issue published online in the open access journal *Water* (ISSN 2073-4441) (available at: https://www.mdpi.com/journal/water/special_issues/wastewater_treatment_valorization_reuse).

For citation purposes, cite each article independently as indicated on the article page online and as indicated below:

LastName, A.A.; LastName, B.B.; LastName, C.C. Article Title. *Journal Name* **Year**, *Volume Number*, Page Range.

ISBN 978-3-0365-1926-5 (Hbk)
ISBN 978-3-0365-1927-2 (PDF)

Contents

About the Editors

Salah Jellali Prof. Salah Jellali has more than 20 years of experience in water and environment research. His research interests include water resources management and governance, solid wastes management, and wastewater treatment, recycling, and reuse. During the last five years, he has focused on biomasses turning into values through the pyrolysis process. He is mainly working on biochars valorization as efficient adsorbents for wastewaters treatment and/or as biofertilizers in agriculture. He has coordinated/participated in several national and international projects related to this theme (GEDURE, BIOCHAR-VAL, SUSTAIN-COAST...). He has published more than 70 peer-reviewed scientific papers and book chapters related to this topic and on the sustainable management of wastewaters in the context of circular economy. Prof. Jellali was a member of the organizing and scientific committees of many international conferences on water and environment, and was also appointed as evaluator of various national and international projects.

Mejdi Jeguirim is a Professor at the University of Haute Alsace (France) in the field of energy and process engineering. He dedicates most of his career to biomass valorization through thermochemical conversion processes including combustion, gasification, and pyrolysis, as well as to the derived chars use in environmental, energy, and agronomy applications. These research topics were performed in the frame of several international collaborations and industrial contracts. He has acted as PhD advisor for 10 students and has co-authored more than 130 referred international journal papers in his research field. He is a member of the editorial board of international journals (Energy, Energies, Energy for Sustainable Development, Biofuels) and the scientific committees of several international congresses. He is involved as a scientific expert for more than 50 international scientific journals, as well as for several national and international research programs. He has received the French National Research Excellence Award for researcher with high level scientific activity for the 2009–2012, 2013–2016, and 2017–2020 periods.

Preface to "Wastewater Treatment, Valorization and Reuse"

Water resources management in general, and wastewater treatment, valorization, and reuse in particular, are currently being considered as important worldwide challenges. In this context, the 2030 United Nations Agenda for Sustainable Development Goals (UN-SDGs) aims to improve water quality by reducing pollution, eliminating dumping, and minimizing the release of hazardous chemicals and materials, halving the proportion of untreated wastewater and substantially increasing recycling and safe reuse globally. Nowadays, the paradigm of wastewater management is shifting from "treatment and disposal" to "reuse, recycle, and resource recovery". The main benefits of this new concept concern not only human and environmental health, food, and energy security, but also climate change mitigation. Hence, in the context of a circular economy, whereby economic development must be balanced with natural resources preservation and environmental sustainability, wastewater can be considered an abundant source of precious and sustainable resources. Therefore, the papers in this Special Issue are dealing with this problematic situation and providing innovative solutions for wastewater treatment and reuse. It is a great pleasure to present this edited volume on wastewater treatment and reuse. It contains 11 valuable publications from esteemed research groups around the world. We would like to thank the editorial team of MDPI, particularly Mrs Mia Liu, for their assistance in this project.

Mejdi Jeguirim, Salah Jellali
Editors

 *water*

Editorial

Wastewater Treatment, Valorization, and Reuse

Mejdi Jeguirim [1,*] **and Salah Jellali** [2]

1   Institut de Science des Matériaux de Mulhouse (IS2M), Université de Strasbourg, Université de Haute-Alsace, CNRS, UMR 7361, F-68100 Mulhouse, France
2   PEIE Research Chair for the Development of Industrial Estates and Free Zones, Center for Environmental Studies and Research (CESAR), Sultan Qaboos University, Al-Khoud, Muscat 123, Oman; s.jellali@squ.edu.om
*   Correspondence: mejdi.jeguirim@uha.fr

Citation: Jeguirim, M.; Jellali, S. Wastewater Treatment, Valorization, and Reuse. *Water* **2021**, *13*, 548. https://doi.org/10.3390/w13040548

Received: 16 January 2021
Accepted: 18 February 2021
Published: 21 February 2021

This Special Issue includes investigations related to wastewater treatment, recovery, and reuse. Different techniques including adsorption, electrocoagulation, ultrafiltration, and membrane filtration were applied to eliminate heavy metals and organic compounds. Furthermore, the use of green technique for water, nutrients and polyhydroxyalkanoates recovery from wastewater was tested. Finally, the potential of the treated wastewater reuse in region under hydric stress was assessed. The different published papers show the diversity of research conducted on wastewater treatment and the environmental and economic benefits from their reuse in various applications. New developments are still in progress, encouraging the organization of another edition of this Special Issue.

Mineral processing wastewater contains large amounts of reagents which can lead to severe environmental problems, such as high chemical oxygen demand (COD). Inspired by the wastewater treatment in such industries as those of textiles, food, and petrochemistry, Jing et al. [1] have applied electrocoagulation (EC) for the first time to explore its feasibility in the treatment of wastewater with an initial COD of 424.29 mg/L from a Pb/Zn sulfide mineral flotation plant and its effect on water reuse. Typical parameters, such as anode materials, current density, initial pH, and additives, were characterized to evaluate the performance of the EC method. The results showed that, under optimal conditions, i.e., iron anode, pH of 7.1, electrolysis time of 70 min, 19.23 mA/cm^2 of current density, and 4.1 g/L of activated carbon dose, the initial COD can be reduced to 72.9 mg/L, corresponding to a removal rate of 82.8%. In addition, compared with the untreated wastewater, EC-treated wastewater was found to benefit the recovery of galena and sphalerite, with galena recovery increasing from 25.01% to 36.06% and sphalerite recovery increasing from 59.99% to 65.33%. This study confirmed that EC is a promising method for the treatment and reuse of high-COD-containing wastewater in the mining industry, and it possesses great potential for a wide range of industrial applications.

Othmani et al. [2] have synthesized hybrid materials with high pollutant-uptake capacity and low cost based on *Luffa cylindrica* (*L.C*) and different percentage of Zn^{2+} in the presence and absence of alternating current (AC). Physico-chemical, morphological, and structural characterizations of the hybrid materials were performed by Boehm method, point zero charge (pH$_{pzc}$), infrared characterizations (IR), scanning electron microscopy (SEM), energy-dispersive spectroscopy, and X-ray photoelectron spectroscopy. The efficiency of the designed hybrid materials was optimized based on their performance in water depollution. Methylene blue (MB) and industrial textile wastewater were the investigated pollutants models. IR characterizations confirmed the fixation of Zn^{2+} onto the *L.C* by the creation of Zn–OH, Zn–O and Zn–O–C bonds. Boehm titration showed that the fixation of Zn^{2+} onto *L.C* is accompanied by an increase in the basic functions of its surface, and subsequently an increase in the pH$_{pzc}$. SEM results confirmed the fixation of Zn^{2+} onto the *L.C*. Coupling AC with biosorption showed an increase in the adsorbed amount of MB and speed. When adding 4% of Zn^{2+} compared to the pure *L.C*, the Q$_m$ shifted from 3.22 to

9.84 and 8.81 mg/g for hybrid materials synthesized under AC, in an absence of AC, and pure *L.C*, respectively.

Miron et al. [3] have studied the evolution of the membrane selectivity of neutral solutes after the filtration of protein or amino acid solutions. Classical methodologies led to the estimation of the mean pore radius, different for each filtrated neutral solute. The use of pore size distribution from nitrogen adsorption/desorption experiments enabled a good description of hydraulic and selectivity performances. The modification of the membrane hydraulic properties after the successive filtration of protein solutions revealed that the decrease is quasi linear, the same for all the studied membranes, and independent of prior tests. According to the experimental observations, an adsorption model was developed, considering a layer-by-layer adsorption in the larger pores of the membrane. The predictive obtained results are in good agreement with the experimental rejection rates, validating the assumptions.

Lignite, as an available and low-cost material, was tested for cadmium (Cd) and copper (Cu) removal from aqueous solutions under various static experimental conditions by Jellali et al. [4]. Experimental results showed that the removal efficiency of both metals was improved by increasing their initial concentrations, adsorbent dosage, and aqueous pH values. The adsorption kinetic was very rapid for Cd; about 78% of the totally adsorbed amounts were removed after a contact time of only 1 min. For Cd and Cu, the kinetic and isothermal data were well fitted with pseudo-second order and Freundlich models, respectively, which suggests that Cd/Cu removal by lignite occurs heterogeneously on multilayer surfaces. The maximum Langmuir's adsorption capacities of Cd and Cu were assessed to 38.0 and 21.4 mg g^{-1} and are relatively important compared to some other lignites and raw natural materials. Results of proximate, scanning electron microscopy/energy dispersive X-ray spectroscopy (SEM/EDS), Fourier-transform infrared spectroscopy (FTIR) and X-ray diffraction (XRD) showed that the removal of these metals occurs most likely through a combination of cation exchange and complexation with specific functional groups. The relatively high adsorption capacity of the used lignite promotes its future use as a low-cost material for Cd and Cu removal from effluents, and possibly for other heavy metals or groups of pollutants.

Greywater has been identified as a potential source of water in a number of applications, e.g., toilet flushing, laundering in first rinsing, floor cleaning, and irrigation. The major obstacle to the reuse of greywater relates to its relatively high contents of pathogens, nutrients, and organic matter. Therefore, much effort has been paid to treat greywater, in order to yield high-quality water deprived of bacteria and with an appropriate value in a wide range of quality parameters (total organic carbon (TOC), nitrate, phosphate, ammonium, pH, and absorbance), similar to the values for tap water. Kamińska and Marszałek [5] proposed to treat real greywater and turn it into high-quality and safe water. For this purpose, the real greywater was treated by means of a sequential biological reactor (SBR) followed by ultrafiltration. Initially, greywater was treated in a laboratory SBR reactor with a capacity of 3 L, operated in a 24 h cycle. Then, SBR effluent was purified in a cross-flow ultrafiltration setup. Treatment efficiency in SBR and ultrafiltration was assessed using extended physicochemical and microbiological analyses (pH, conductivity, color, absorbance, chemical oxygen demand (COD), biological oxygen demand (BOD$_5$), nitrate, phosphate, ammonium, total nitrogen, phenol index, nonionic and anionic surfactants, TOC, *Escherichia coli*, and enterococci). Additionally, ultrafiltration was evaluated in terms of fouling behavior for three polymer membranes with different MWCOs (molecular weight cut-offs). The values of quality parameters (pH, conductivity, COD, BOD$_5$, TOC, N-NH$_4^+$, N-NO$_3^-$, N$_{tot}$, and P-PO$_4^{3-}$) measured in SBR effluent did not exceed permissible values for wastewater discharged to soil and water. Ultrafiltration provided the high-quality water with very low values of COD (5.8–18.1 mg/L), TOC (0.47–2.19 mg/L), absorbance UV254 (0.015–0.048 1/cm), color (10–29 mgPt/L), and concentration of nitrates (0.18–0.56 mg/L), phosphates (0.9–2.1 mg/L), ammonium (0.03–0.11 mg/L), and total nitrogen (3.3–4.7 mg/L), as well as a lack of *E. coli* and enterococci. Membrane structural and surface properties did not affect the treatment efficiency, but did influence the fouling behavior.

Water shortage is a very concerning issue in the Mediterranean region, threatening the viability of the agriculture sector, and in some countries, population wellbeing. At the same time, liquid effluent volumes generated from agro-food industries in general, and the olive oil industry in particular, are huge. Thus, Dutournié et al. [6] proposed a sustainable solution for the management of olive mill wastewater (OMWW) with possible reuse in irrigation. Their investigation is a part of a series of papers valorizing all the outputs of a three-phase system of olive oil mills. It deals with recovery, by condensation, of water from both OMWW and OMWW-impregnated biomasses (sawdust and wood chips), during a convective drying operation (air velocity: 1 m/s and air temperature: 50 °C). The experimental results showed that the water yield recovery reaches about 95%. The condensate waters have low electrical conductivity and salinities, but also acidic pH values and slightly high chemical oxygen demand (COD) values. However, they could be returned suitable for reuse in agriculture after additional low-cost treatments.

Raw poultry manure (RPM) and its derived biochars at temperatures of 400 (B400) and 600 °C (B600) were physico-chemically characterized, and their ability to release nutrients was assessed under static conditions by Hadroug et al. [7]. The experimental results showed that RPM pyrolysis operation significantly affects its morphology, surface charges, and area, as well as its functional groups contents, which in turn influences its nutrient release ability. The batch experiments indicated that nutrient release from the RPM as well as biochars attains a pseudo-equilibrium state after a contact time of about 48 h. RPM pyrolysis increased phosphorus stability in residual biochars and, in contrast, transformed potassium to a more leachable form. For instance, at this contact time, P- and K-released amounts changed from 5.1 and 25.6 mg g^{-1} for RPM to only 3.8 and more than 43.3 mg g^{-1} for B400, respectively. On the other hand, six successive leaching batch experiments with a duration of 48 h each showed that P and K release from the produced biochars was a very slow process; negligible amounts continued to be released even after a total duration of 12 days. All these results suggest that RPM-derived biochars have specific physico-chemical characteristics, allowing them to be used in agriculture as low-cost and slow-release fertilizers.

Sludge from municipal wastewater treatment systems can be used as a source of mixed microbial cultures for the production of polyhydroxyalkanoates (PHA), according to de Souza Reis et al. [8]. Stored intracellularly, the PHA is accumulated by some species of bacteria as an energy stockpile and can be extracted from the cells by reflux extraction. Dimethyl carbonate was tested as a solvent for the PHA extraction at different extraction times and biomass-to-solvent ratios, and 1-butanol was tested for purifying the obtained PHA at different purification times and PHA-to-solvent ratios. Overall, only a very small difference was observed in the different extraction scenarios. An average extraction amount of 30.7 ± 1.6 g of PHA per 100 g of biomass was achieved. After purification with 1-butanol, a visual difference was observed in the PHA between the tested scenarios, although the actual purity of the resulting samples did not present a significant difference. The overall purity increased from 91.2 ± 0.1% to 98.0 ± 0.1%.

Current regulations and legislation require critical revision to determine safety for alternative water sources and water reuse as part of the solution to the global water crisis. In order to fulfill those demands, the Lisbon municipality decided to start water reuse as part of a sustainable hydric resource management, and there was a need to confirm safety and safeguard for public health for its use in this context. For this purpose, a study was designed that included a total of 88 samples collected from drinking, superficial, underground water, and wastewater at three different treatment stages [9]. Quantitative polymerase chain reaction (PCR) detection (qPCR) of enteric viruses norovirus (NoV) genogroups I (GI) and II (GII) and hepatitis A (HepA) was performed, and FIB (*E. coli*, enterococci and fecal coliforms) concentrations were also assessed. HepA virus was only detected in one untreated influent sample, whereas NoV GI/NoV GI were detected in untreated wastewater (100/100%), secondary treated effluent (47/73%), and tertiary treated effluent (33/20%). This study proposes that NoV GI and GII should be further studied to provide the support that they may be suitable indicators for water quality monitoring targeting wastewater treatment efficiency, regardless of the level of treatment.

Water scarcity remains the major looming challenge which is facing Jordan. Wastewater reclamation is considered as an alternative source of fresh water in semi-arid areas with water shortage or increased consumption. The current status of wastewater reclamation and reuse in Jordan was analyzed considering 30 wastewater treatment plants (WWTPs) [10]. The assessment was based on the WWWTPs' treatment processes in Jordan, the flowrate scale, and the effluents' average total dissolved solid (TDS) contents. Accordingly, 60% of the WWTPs in Jordan used activated sludge as a treatment technology; 30 WWTPs were small scale ($<1 \times 10^4$ m^3/day); and a total of 17.932 million m^3 treated wastewater had low TDS (<1000 ppm) that generally can be used in industries with relatively minimal cost of treatment. Moreover, the analysis classified the 26 million m^3 groundwater abstraction by major industries in Jordanian governorates. The results showed that the reclaimed wastewater can fully offset the industrial demand of fresh water in the Amman, Zarqa, and Aqaba governorates. Hence, the environmental assessment showed positive impacts of reclaimed wastewater reuse scenario in terms of water depletion (saving of 72.55 million m^3 groundwater per year) and climate change (17.683 million kg CO_{2Eq} reduction). The energy recovery assessment in the small- and medium-scale WWTPs ($<10 \times 10^4$ m^3/day) revealed that generation of electricity by anaerobic sludge digestion equates potentially to an offset of 0.11–0.53 kWh/m^3. Finally, several barriers and prospects were imposed to help the stakeholders when considering entering into an agreement to supply and/or reuse reclaimed water.

We found the edition and selections of papers for this Special Issue very inspiring and rewarding. We also thank the editorial staff and reviewers for their efforts and help during the process.

Conflicts of Interest: The authors declare no conflict of interest.

References

1. Jing, G.; Ren, S.; Gao, Y.; Sun, W.; Gao, Z. Electrocoagulation: A Promising Method to Treat and Reuse Mineral Processing Wastewater with High COD. *Water* **2020**, *12*, 595. [CrossRef]
2. Othmani, A.; Kesraoui, A.; Boada, R.; Seffen, M.; Valiente, M. Textile Wastewater Purification Using an Elaborated Biosorbent Hybrid Material (*Luffa–Cylindrica*–Zinc Oxide) Assisted by Alternating Current. *Water* **2019**, *11*, 1326. [CrossRef]
3. Miron, S.M.; Dutournié, P.; Ponche, A. Filtration of Uncharged Solutes: An Assessment of Steric Effect by Transport and Adsorption Modelling. *Water* **2019**, *11*, 2173. [CrossRef]
4. Jellali, S.; Azzaz, A.A.; Jeguirim, M.; Hamdi, H.; Mlayah, A. Use of Lignite as a Low-Cost Material for Cadmium and Copper Removal from Aqueous Solutions: Assessment of Adsorption Characteristics and Exploration of Involved Mechanisms. *Water* **2021**, *13*, 164. [CrossRef]
5. Kamińska, G.; Marszałek, A. Advanced Treatment of Real Grey Water by SBR Followed by Ultrafiltration—Performance and Fouling Behavior. *Water* **2020**, *12*, 154. [CrossRef]
6. Dutournié, P.; Jeguirim, M.; Khiari, B.; Goddard, M.-L.; Jellali, S. Olive Mill Wastewater: From a Pollutant to Green Fuels, Agricultural Water Source, and Bio-Fertilizer. Part 2: Water Recovery. *Water* **2019**, *11*, 768. [CrossRef]
7. Hadroug, S.; Jellali, S.; Leahy, J.J.; Kwapinska, M.; Jeguirim, M.; Hamdi, H.; Kwapinski, W. Pyrolysis Process as a Sustainable Management Option of Poultry Manure: Characterization of the Derived Biochars and Assessment of their Nutrient Release Capacities. *Water* **2019**, *11*, 2271. [CrossRef]
8. De Souza Reis, G.A.; Michels, M.H.A.; Fajardo, G.L.; Lamot, I.; de Best, J.H. Optimization of Green Extraction and Purification of PHA Produced by Mixed Microbial Cultures from Sludge. *Water* **2020**, *12*, 1185. [CrossRef]
9. Teixeira, P.; Costa, S.; Brown, B.; Silva, S.; Rodrigues, R.; Valério, E. Quantitative PCR Detection of Enteric Viruses in Wastewater and Environmental Water Sources by the Lisbon Municipality: A Case Study. *Water* **2020**, *12*, 544. [CrossRef]
10. Saidan, M.N.; Al-Addous, M.; Al-Weshah, R.A.; Obada, I.; Alkasrawi, M.; Barbana, N. Wastewater Reclamation in Major Jordanian Industries: A Viable Component of a Circular Economy. *Water* **2020**, *12*, 1276. [CrossRef]

 *water*

Article

Electrocoagulation: A Promising Method to Treat and Reuse Mineral Processing Wastewater with High COD

Gaogui Jing [1,2], Shuai Ren [1,2], Yuesheng Gao [3], Wei Sun [1,2] and Zhiyong Gao [1,2,*]

[1] School of Minerals Processing and Bioengineering, Central South University, Changsha 410083, China; gaoguijing@csu.edu.cn (G.J.); renshuai@csu.edu.cn (S.R.); sunmenghu@126.com (W.S.)

[2] Key Laboratory of Hunan Province for Clean and Efficient Utilization of Strategic Calcium-Containing Mineral Resources, Central South University, Changsha 410083, China

[3] Department of Chemical Engineering, Michigan Technological University, Houghton, MI 49931, USA; ygao4@mtu.edu

*   Correspondence: zhiyong.gao@csu.edu.cn

Received: 23 January 2020; Accepted: 18 February 2020; Published: 21 February 2020

Abstract: Mineral processing wastewater contains large amounts of reagents which can lead to severe environmental problems, such as high chemical oxygen demand (COD). Inspired by the wastewater treatment in such industries as those of textiles, food, and petrochemistry, in the present work, electrocoagulation (EC) is applied for the first time to explore its feasibility in the treatment of wastewater with an initial COD of 424.29 mg/L from a Pb/Zn sulfide mineral flotation plant and its effect on water reuse. Typical parameters, such as anode materials, current density, initial pH, and additives, were characterized to evaluate the performance of the EC method. The results showed that, under optimal conditions, i.e., iron anode, pH 7.1, electrolysis time 70 min, 19.23 mA/cm^2 current density, and 4.1 g/L activated carbon, the initial COD can be reduced to 72.9 mg/L, corresponding to a removal rate of 82.8%. In addition, compared with the untreated wastewater, EC-treated wastewater was found to benefit the recovery of galena and sphalerite, with galena recovery increasing from 25.01% to 36.06% and sphalerite recovery increasing from 59.99% to 65.33%. This study confirmed that EC is a promising method for the treatment and reuse of high-COD-containing wastewater in the mining industry, and it possesses great potential for wide industrial applications.

Keywords: mineral processing; wastewater treatment; flotation; electrocoagulation (EC); chemical oxygen demand (COD)

1. Introduction

Wastewater from mineral processing remains a headache for the mining industry not only because of its large volume, but more importantly, its hazardous components. Among the complicated residual reagents, the organic reagents receive the most attention due to their high chemical oxygen demand (COD) levels. Direct discharge of wastewater causes serious environmental pollution, while its reuse as feed water without proper treatments causes great harm to the recovery and grade of valuable minerals [1]. To solve this problem, techniques for wastewater treatment have been intensively developed in the mining industry, such as natural settling, flocculating setting, chemical oxidation, adsorption, and biodegradation to remove COD [2]. However, these techniques possess such disadvantages as inefficiency in COD removal, extreme operating conditions, high operation cost, huge instrument investment, etc. It is worth noting that coagulation–flocculation, the most common technique in beneficiation plants because of its economy, usually has a relative low COD removal rate of only 35% [3].

Electrocoagulation (EC) has become widely used as a wastewater treatment technology in the past decade and has exhibited a satisfactory removal of COD in diverse fields, such as in textiles [4], the food industry [5], dairying [6], brackish water [7], and potable water [8]. Its advantages include easy operation, small area occupation, and a high degree of automation. In the process of EC, iron or aluminum is usually adopted as the anode material, and each one can lead to a series of similar redox reactions. Taking the iron electrode as an example, simplified reactions occurring in acidic and alkaline solutions can be described as follows [9]:

Under acidic conditions:

$$Fe_{(s)} \rightarrow Fe^{2+}_{(aq)} + 2e^- \tag{1}$$

$$2H_2O \rightarrow O_2 + 4H^+ + 4e^- \tag{2}$$

$$4Fe^{2+}_{(aq)} + 10H_2O_{(l)} + O_{2(g)} \rightarrow 4Fe(OH)_{3(s)} + 8H^+_{(aq)} \tag{3}$$

Under alkaline conditions:

$$Fe_{(s)} \rightarrow Fe^{2+}_{(aq)} + 2e^- \tag{4}$$

$$Fe^{2+} + 2OH^- \rightarrow Fe(OH)_2 \tag{5}$$

As shown in above equations, the iron anode is used to generate ions in the aqueous medium. The generated iron ions are then immediately hydrolyzed to form $Fe(OH)_3$, monomeric ions, and polymeric hydroxy complexes such as $Fe(H_2O)_6^{3+}$, $Fe(H_2O)_4(OH)^{2+}$, $Fe_2(H_2O)_6(OH)_4^{4+}$, $Fe_2(H_2O)_8(OH)_2^{4+}$, and $Fe_2(H_2O)_5(OH)^{2+}$, depending on the solution pH [10]. These hydroxides and polyhydroxides are excellent adsorbents for counter ions and organic pollutants. In some cases, they can even form complexes with some of the organics [11]. Pollutants in the wastewater can be treated by physical and chemical attachment to hydroxides or chemical reactions [12]. After the EC treatment, separation of precipitation from the liquid is traditionally achieved via a filter process. Then, the filtered and separated sludge is usually treated using the incineration and landfill treatments [13].

In this paper, EC treatment of sulfide mineral flotation wastewater was applied for the first time, and the effects of different parameters like anode materials, current density, initial pH, electrolysis time, and additives on COD removal rate were investigated. The treated water was then recycled as feed water in the sulfide ore flotation. For comparison, untreated wastewater and fresh water were used to assess the effects of EC-treated water on the grade and recovery of galena and sphalerite.

2. Materials and Methods

2.1. Samples, Reagents, and Analytical Methods

Samples of mineral processing wastewater and actual ores were all from a Pb-Zn beneficiation plant located in Southern China. The plant had three different types of wastewater, namely Pb-Zn flotation wastewater, pyrite concentrate wastewater, and pyrite tailing wastewater, accounting for 70%, 14%, and 16% of the total amount, respectively. In this study, three different types of wastewater samples were mixed at the above ratio to obtain the mixed wastewater samples. The pH and COD values of each wastewater are shown in Table 1.

Table 1. Chemical oxygen demand (COD) and pH values of mineral processing wastewater samples in a Pb-Zn beneficiation plant.

Sample Name	pH	COD (mg/L)
Pb-Zn flotation wastewater	11.90	434.7
Pyrite concentrate wastewater	9.11	245.3
Pyrite tailing wastewater	8.29	603.2
Mixed wastewater	11.83	424.9

The H$_2$O$_2$ (30% pure), H$_2$SO$_4$, HCl, and NaOH used in the EC experiments were all of analytical grades. The granular activated carbon (GAC) was made from a coconut shell.

The pH of the solution was measured by a pH meter (Leici, Shanghai, China). The COD values of the wastewater were analyzed by fast digestion–spectrophotometric method (HJ/T 399-2007).

The chemical compositions of the actual ore samples were relatively simple. SiO$_2$ and CaO were the main non-metallic ingredients [14]. The lead and zinc grades in the samples took up 4.17% and 7.18% of the total, respectively. The major metallic minerals were galena, sphalerite, and pyrite. Actual ore samples were crushed to below 3 mm using a stainless-steel hammer and then ground in a conical ball mill. The ground products were used as feed samples for flotation tests.

Diethyldithiocarbamate (95% purity) and sodium n-butylxanthate (92% purity) were purchased from Aladdin Industrial, Shanghai, China. The CaO (analytically pure) and CuSO$_4$ (analytically pure) used in this study were provided by Kermel Company, Tianjin, China. Terpenic oil (chemically pure) was obtained from Haoshen Chemical Reagent, Shanghai, China.

2.2. EC Experiments

The EC set-up used in this study is shown in Figure 1. The total volume of the electrolytic cell was 250 mL, and the experiments were carried out by treating 200 mL wastewater samples for each run at room temperature. Two pairs of plates (length = 6.5 cm; width = 5.6 cm; thickness = 0.2 cm) were placed on the electrolytic cell in parallel. The anode and cathode plates were connected to the positive and negative ports of the QW-MS3010D DC power supply (0–5 A, 0–80 V), with constant load current. NaOH and H$_2$SO$_4$ solutions were used to adjust the pH. At the end of the experiment, water was transferred to beakers for two hours of settling, and then the COD value of the supernatant was measured. Finally, the sludge produced by the EC was filtered, dried, and weighed.

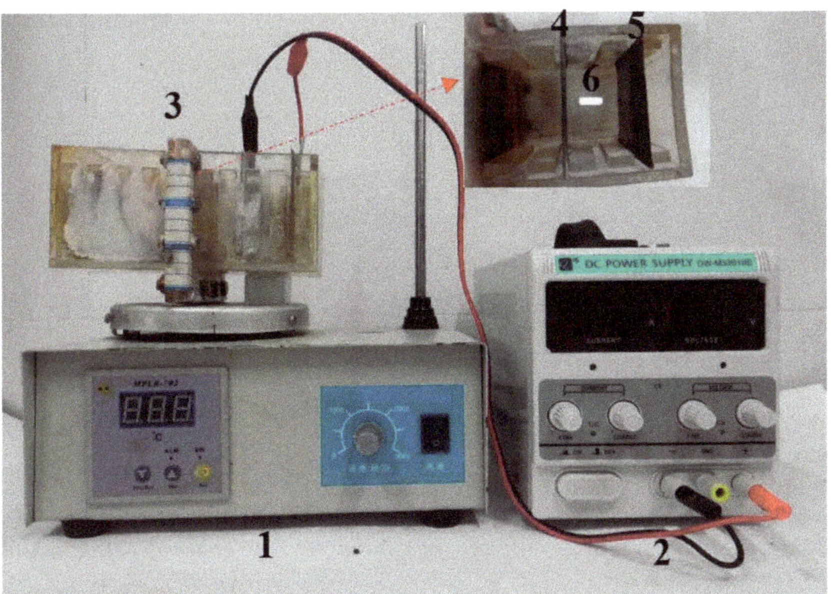

Figure 1. Electrocoagulation (EC) experimental set-up: (1) Magnetic stirrer; (2) DC power supply; (3) electrolytic cell; (4) anode plate; (5) cathode plate; (6) magnetic bar-stirrer.

The COD removal rate was employed to evaluate the electrocoagulation efficiency, which is calculated according to the following Equation (6).

$$R(\%) = \frac{COD_0 - COD_1}{COD_0} \times 100,$$ (6)

where R represents the removal rate of COD (%), COD_0 represents the initial COD value of the wastewater before electrocoagulation (mg/L), and COD_1 represents the COD value of the wastewater after electrocoagulation (mg/L).

2.3. Flotation Experiments

Fresh water, wastewater, and treated water were then used for flotation at room temperature. The whole process can be divided into two stages: Galena flotation followed by sphalerite flotation. For each flotation experiment, an 800 g ore sample was ground by a conical ball mill to obtain a product of 85 wt.% passing 74 μm. After grinding, the ore pulp was transferred to the galena flotation stage, including single-stage roughing, three-stage cleaning, and single-stage scavenging. The tailing of galena flotation was used later for sphalerite flotation, where single-stage roughing, two-stage cleaning, and single-stage scavenging were adopted. In both flotation stages, the concentrates of roughing were reground to 80 wt.% passing 44 μm in a Φ150 × 100 conical mill. According to the existing process of the beneficiation plant, sequential addition of regulator, collector, and frother to the flotation cell were carried out, as shown in Figure 2. After flotation, the products were dried, weighed, and assayed, and then the recovery was calculated on the basis of the mass and metal balance, as shown in Equation (7).

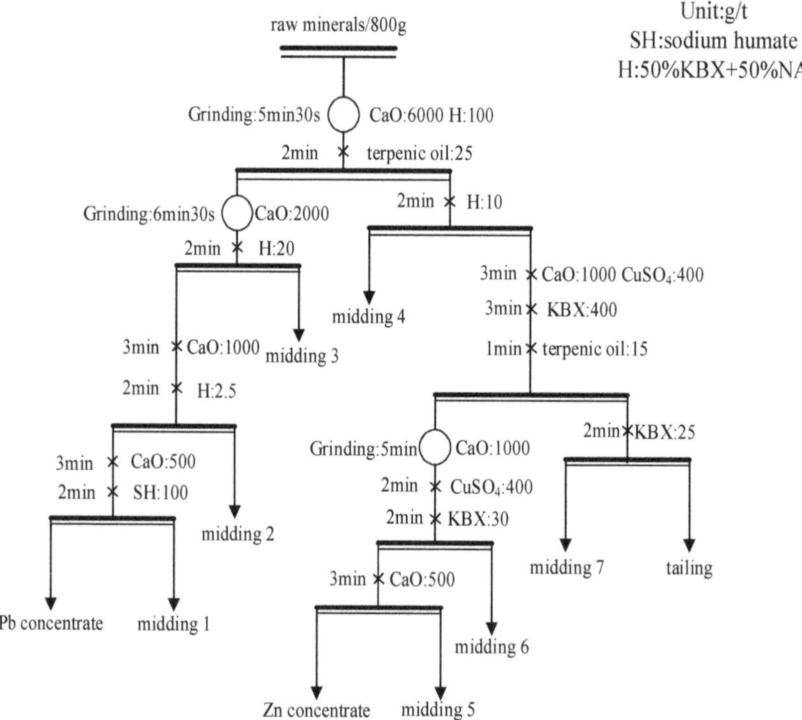

Figure 2. Flowsheet and corresponding experimental conditions of batch flotation tests using different types of water.

$$\varepsilon(\%) = \frac{\gamma_i \times \beta_i}{\sum_1^n (\gamma_i \times \beta_i)} \times 100, \tag{7}$$

where ε is the recovery of product, γ_i is the yield, and β_i is the grade.

3. Results and Discussion

3.1. Effect of EC Parameters on COD Removal Rate of Mixed Wastewater

Unlike anode materials, cathode materials do not have a significant effect on EC efficiency except in electro-Fenton processes [15]. We have learned that stainless steel is the most widely-used cathode material in industry, considering its economy and durability [16]. In addition, the distance between the anode and cathode is suggested to be 1~2 cm, according to previous publications and industrial experience [17]. After verification, 2 cm was the optimal distance to be adopted in our following experiments. As the anode material is a prime consideration for EC technique [18], iron and aluminum plates were used as anode materials in this study. The results of COD removal efficiency with the two anode materials are shown in Figure 3. When using an iron electrode, the COD value decreases from 424.9 to 158.6 mg/L, which is equivalent to a removal rate of 62.7%. However, when an aluminum anode is used, the COD value is reduced to 265.1 mg/L, corresponding to a removal rate of 37.7%. The results indicate that in this study, Fe performs better as the anode material of the EC process. Therefore, an Fe anode was adopted in the subsequent experiments.

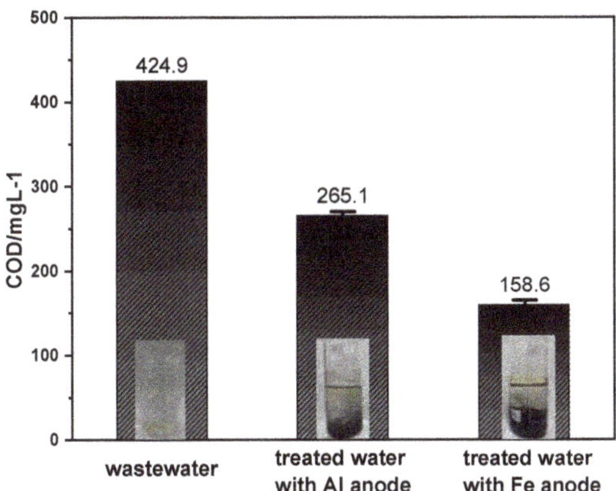

Figure 3. Effect of anode materials for EC treatment on mixed wastewater COD value (cathode material: Stainless steel, current density: 13.74 mA/cm^2, pH: 7.1, electrolysis time: 50 min).

Current density is another important parameter that determines the amount of coagulant produced in the reactor [19]. It has a significant effect on the EC efficiency and electricity consumption. The experimental results are summarized in Figure 4. Figure 4 shows that COD removal rate first increases and then declines with the increase of current density. COD removal rate increases from 49.41% to 77.46% as the current density rises to 24.73 mA/cm^2. Then, it decreases to 60.42% as the current density continues to increase to 30.22 mA/cm^2. The current density is related to the amount of dissolved iron ions and the formation of Fe(OH)$_3$, which determines the COD removal efficiency [20]. It can also affect the cell voltage through various overpotentials, which may be conducive to the removal efficiency of the COD. If the current density is relatively low, the lack of coagulant may restrict the removal efficiency. Therefore, the curve presents an upward trend at an early stage. However, when the current density continues to increase to 32.22 mA/cm^2, the temperature of treated wastewater increases significantly. High temperature causes the instability of coagulation, which can lead to a low EC efficiency. Furthermore, energy consumption grows gradually with the increase of current density [21]. Taking both the COD removal rate and electricity consumption into account, 19.23 mA/cm^2 was used as the current density for subsequent experiments.

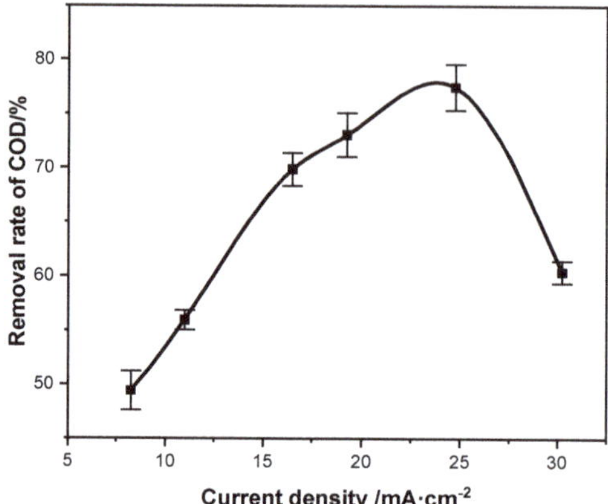

Figure 4. Effect of current density on mixed wastewater COD removal rate (anode material: Fe, cathode material: Stainless steel, pH: 7.1, electrolysis time: 50 min).

To evaluate the effect of the initial pH, the wastewater samples were adjusted to a preset pH value using sodium hydroxide or sulfuric acid. The effects of initial pH value on COD removal rate are shown in Figure 5. The results show that the COD removal rate in the EC process remains around 75.0% in a wide pH value range from 3 to 9. However, when the initial pH increases to 11, the removal efficiency presents a downward trend with a corresponding removal rate of 71.08%; this is probably because of the formation of soluble $Fe(OH)^{-4}$, which is not conducive to the floc formation. On the contrary, at a neutral or acidic pH, higher removal rates occur as most of the iron complexes are formed. Hence, subsequent experiments were carried out under neutral pH conditions. pH 7.1 was employed in the following experiments.

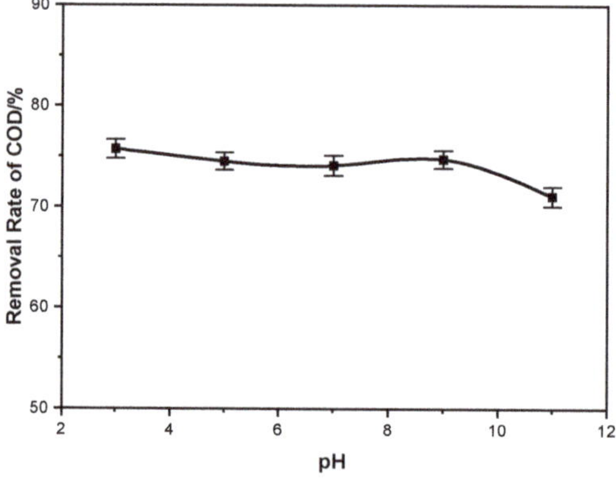

Figure 5. Effect of initial pH value on mixed wastewater COD removal rate (anode material: Fe, cathode material: Stainless steel, current density: 19.23 mA/cm^2, electrolysis time: 50 min).

Electrolysis time is another important parameter affecting COD removal efficiency [22]. The results of the effect of electrolysis time on the COD removal rate are provided in Figure 6. Figure 6 shows that the COD removal rate increases rapidly in the initial phase, which is due to the increase of coagulant;

yet, fe growth trend of the COD removal rate gradually slows down with the extension of electrolysis time. When electrolysis time increases to 70 min, the COD removal rate reaches a plateau at 77.62% because the adsorption of organic pollutants reaches an equilibrium state. The electrolysis time of subsequent experiments was set to be 70 min.

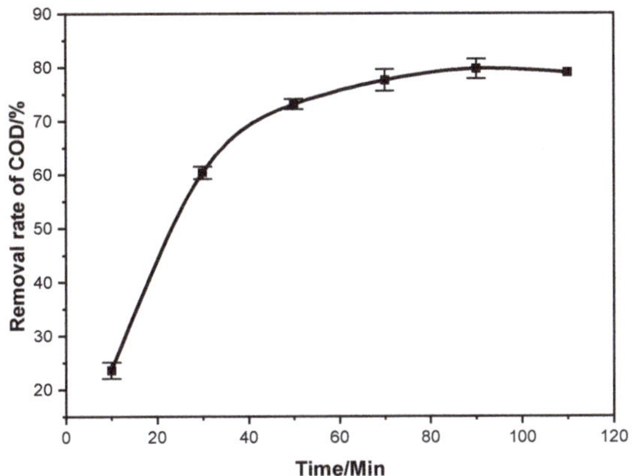

Figure 6. Effect of electrolysis time on mixed wastewater COD removal rate (anode material: Fe, cathode material: Stainless steel, current density: 19.23 mA/cm^2, pH: 7.1).

3.2. Effect of Additives on COD Removal Rate of Mixed Wastewater

Under the same current density, an increase of supporting electrolyte concentration can lead to a decrease of the interelectrode resistance and thus reduce energy consumption. Ghernaout reported that the addition of a supporting electrolyte can also affect the removal rate of COD [23]. Hence, the effect of supporting electrolyte concentration on EC efficiency was further studied in this study, and Na_2SO_4 was chosen as the supporting electrolyte because of its low cost. The Na_2SO_4 concentration of 1.0, 1.1, 1.2, and 1.3 g/L corresponded to the current densities of 8.24, 13.74, 19.23, and 24.73 mA/cm^2, respectively. The results of the effect of Na_2SO_4 addition on the COD removal rate are shown in Figure 7. The removal rates with the addition of Na_2SO_4 are much lower than those without Na_2SO_4. No positive effect on COD removal efficiency is observed by the addition of Na_2SO_4, which is consistent with previous reports [24]. The low removal efficiency can be attributed to the increase of the passivation layer caused by the addition of Na_2SO_4 [25].

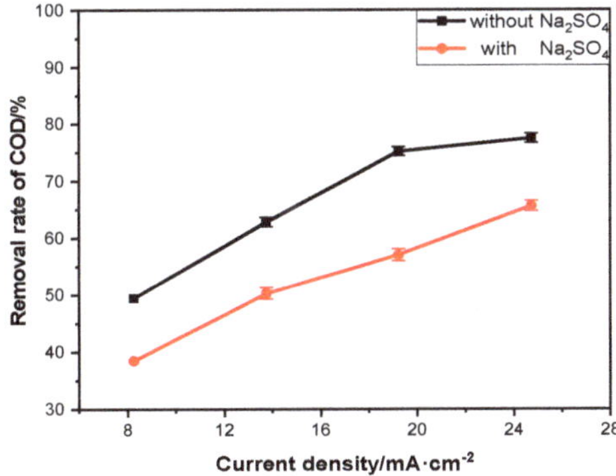

Figure 7. Effect of addition of Na_2SO_4 on mixed wastewater COD removal rate (anode material: Fe, cathode material: Stainless steel, pH: 7.1, electrolysis time: 70 min).

Activated carbon is widely used to adsorb organic compounds in wastewater [26]. In order to study the effect of activated carbon on the COD removal efficiency with EC, wastewater samples were treated by GAC adsorption, EC/GAC adsorption (EC plus GAC in one reactor), and EC + GAC adsorption (EC followed by GAC in different reactors) process, and their results are summarized in Figure 8.

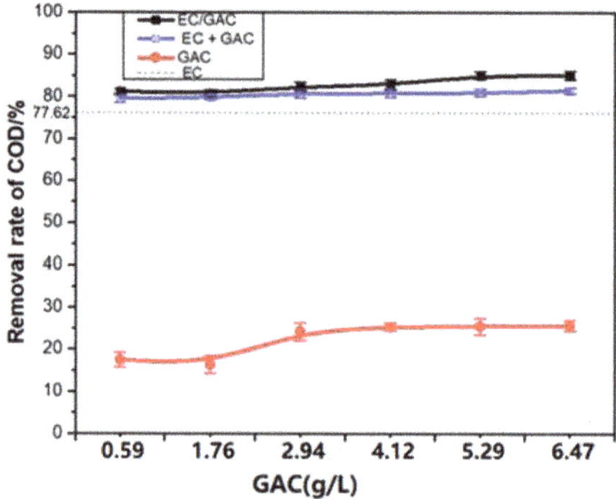

Figure 8. Effect of granular activated carbon (GAC) treatment on mixed wastewater COD removal rate (anode material: Fe, cathode material: Stainless steel, current density: 19.27 mA/cm^2, pH: 7.1, electrolysis time: 70 min).

The Figure 8 shows that the COD removal rate using GAC alone is low (only 25.7%) when the GAC dosage reaches 6.47 g/L, indicating that GAC adsorption has little effect on the residual reagents in mineral processing wastewater. When wastewater is treated by the EC process alone, the COD removal rate is 77.62%. Compared with EC or GAC alone, both the EC/GAC and EC + GAC processes can improve COD removal. However, it is worth noting that the COD removal rate of EC/GAC is higher than that of EC + GAC. At a dosage of 2.94 g/L GAC, the COD decreases from 424.9 to 74.9 and 83.3 mg/L, corresponding to removal rates of 82.37% for EC/GAC and 80.4% for EC/GAC, respectively. Therefore, the EC/GAC process was adopted in the subsequent experiments.

It can be concluded that the adsorption of GAC alone cannot be used as an effective treatment method for sulfide mineral flotation wastewater. In the EC/GAC system, GAC acts as a moving particle electrode to accelerate electron transfer rate and hence improve the processing efficiency [27,28]. The main reason is that it is essentially a three-dimensional electrochemical process which is not stable [29].

The sludge produced is around 5.8 kg/m^3 of wastewater under the optimal processing conditions. It can be considered as a valuable resource because of its high Fe content of about 45.5%. Further research on the sludge recycling utilization will soon be reported separately.

Furthermore, the effect of H$_2$O$_2$ concentration on EC efficiency was studied, and the results can be seen in the Supplementary Materials (Figure S1 and Figure S2).

3.3. Effect of EC Treatment on COD Removal Rate of Different Types of Wastewater

Through the above experiments, the optimum treatment conditions of mixed wastewater were obtained (iron anode, pH 7.1, 70 min electrolysis time, 19.23 mA/cm^2 current density, and 4.1 g/L activated carbon). Under these optimum conditions, Pb-Zn flotation wastewater, pyrite concentrate wastewater, pyrite tailing wastewater, and mixed wastewater were treated, respectively. The results are shown in Figure 9.

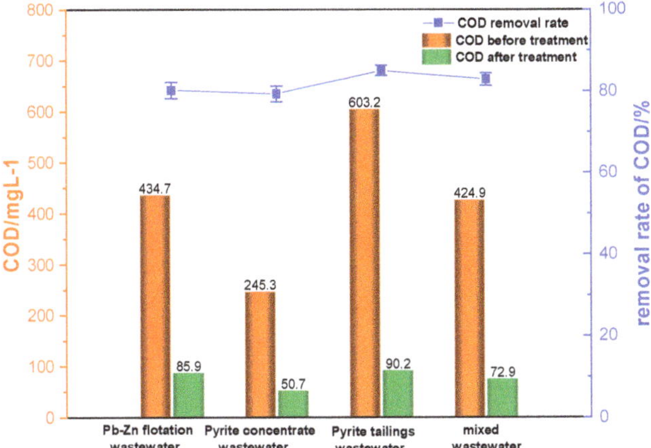

Figure 9. Effect of EC treatment on the COD values and COD removal rates of four different types of wastewaters.

The COD removal rates for the above four types of wastewaters are 80.24%, 79.33%, 85.05%, and 82.84%, respectively. The results indicate that the higher the COD value of wastewater is, the better efficiency the corresponding EC experiment obtains. The pyrite tailing wastewater with the highest COD value has the highest COD removal rate, followed by mixed wastewater, Pb-Zn flotation wastewater, and pyrite concentrate wastewater.

3.4. Effect of Water Type on the Grade and Recovery of Pb/Zn Sulfide Mineral Flotation

The above experimental results confirmed that EC is an effective technique for COD removal of the sulfide mineral flotation wastewater. To further investigate the possibility of reusing the EC-treated water as the feed water, flotation experiments were carried out, and their results are shown in Tables S1–S3, respectively. The comparison of the Pb/Zn grades and recovery rates using different types of water during flotation is presented in Figure 10.

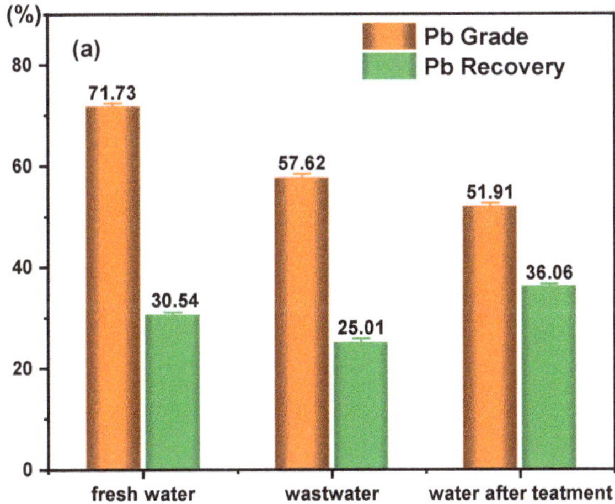

Figure 10. *Cont.*

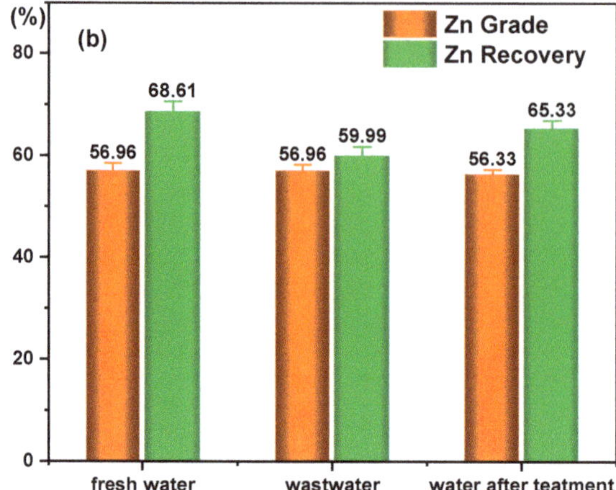

Figure 10. The grade and recovery of Pb (**a**) and Zn (**b**) sulfide flotation concentrates using different types of water.

The Pb recovery using different types of water obeys the following order: Treated water (36.06%) > fresh water (30.54%) > wastewater (25.01%), whereas the order of Pb grades is: Fresh water (71.73%) > wastewater (57.62%) > treated water (51.91%). The experimental results indicate that flotation using treated water can significantly improve the Pb recovery, while the Pb grade is reduced.

In terms of Zn concentrate, a similar Zn grade is obtained with different types of waters (around 56%). The Zn recovery obeys the following order: Fresh water (68.61%) > treated water (65.33%) > wastewater (59.99%). Compared with the untreated wastewater, EC-treated water is more conducive for the increase of Zn recovery without affecting the Zn grade.

4. Conclusions

This study provides the latest evidence that EC is a promising method for treating and recycling mineral processing wastewater. Different factors were investigated to obtain optimal experimental conditions, i.e., iron anode, pH 7.1, 19.23 mA/cm^2 current density, 70 min electrolysis time, and 4.1 g/L activated carbon. Under the optimal conditions, the COD of the mixed wastewater decreases from 424.9 to 72.9 mg/L. The addition of a supporting electrolyte like Na_2SO_4 can be harmful to the removal of COD. The COD removal rate of wastewater with the highest COD value is highest. Moreover, flotation results show that the recoveries of both Pb and Zn using EC-treated water are higher than those of directly using wastewater or fresh water. The results imply that the EC technique can provide a new direction for recycling mineral processing wastewater. This study provides novel insights into dealing with wastewater in the mining industry, and is meaningful in both environmental and economical ways.

Supplementary Materials: The following are available online at http://www.mdpi.com/2073-4441/12/2/595/s1: Figure S1: Variation of COD removal rate with dosage of H_2O_2 (Anode material: Fe, Cathode material: Stainless steel, Current density: 19.23 mA/cm^2, pH: 3.0, Electrolysis time: 70 min), Figure S2: Effect of H_2O_2 on the boundary between mud and water, Table S1: Flotation tests results with fresh water, Table S2: Flotation tests results with mixed wastewater, Table S3: Flotation tests results with treated water.

Author Contributions: Conceptualization, Z.G. and W.S.; methodology, W.S.; formal analysis, Z.G.; investigation, G.J.; data curation, G.J. and S.R.; writing—original draft preparation, G.J.; writing—review and editing, Y.G.; supervision, Z.G.; project administration, W.S.; funding acquisition, Z.G. All authors have read and agreed to the published version of the manuscript.

Funding: The authors acknowledge financial support from the National Natural Science Foundation of China (51634009, 51774328, and 51404300), the Young Elite Scientists Sponsorship Program by CAST (2017QNRC001), the Young Elite Scientists Sponsorship Program by Hunan province of China (2018RS3011), the Natural Science Foundation of Hunan Province of China (2018JJ2520), and the National 111 Project (B14034).

Acknowledgments: The authors are grateful to the financial support from the National Natural Science Foundation of China (51634009, 51774328, and 51404300), the Young Elite Scientists Sponsorship Program by CAST (2017QNRC001), the Young Elite Scientists Sponsorship Program by Hunan province of China (2018RS3011), the Natural Science Foundation of Hunan Province of China (2018JJ2520), and the National 111 Project (B14034). All authors appreciate the School of Minerals Processing and Bioengineering, Central South University, Changsha, China for providing the facilities.

Conflicts of Interest: The authors declare no conflict of interest.

References

1. Farrokhpay, S.; Zanin, M. An investigation into the effect of water quality on froth stability. *Adv. Powder Technol.* **2012**, *23*, 493–497. [CrossRef]
2. Zhao, S.; Pan, J.Z. Overview on the Treatment Technology of Wastewater from Copper Mineral Processing (WCMP). *Adv. Mater. Res.* **2013**, *750*, 1369–1372.
3. Sitorus, F.; Cilliers, J.J.; Brito-Parada, P.R. Multi-criteria decision making for the choice problem in mining and mineral processing: Applications and trends. *Expert Syst. Appl.* **2019**, *121*, 393–417. [CrossRef]
4. Kobya, M.; Can, O.T.; Bayramoglu, M. Treatment of textile wastewaters by electrocoagulation using iron and aluminum electrodes. *J. Hazard. Mater.* **2003**, *99*, 163–178. [CrossRef]
5. Beck, E.C.; Giannini, A.P.; Ramirez, E.R. Electrocoagulation clarifies food wastewater. *Food Technol.* **1974**, *22*, 18–19.
6. Bazrafshan, E. Application of electrocoagulation process for dairy wastewater treatment. *J. Chem.* **2012**, *2013*, 1–8. [CrossRef]
7. Wang, S.; Wang, Q. Experimental studies on pretreatment process of brackish water using electrocoagulation (EC) method. *Desalination* **1987**, *66*, 353–364. [CrossRef]
8. Vik, E.A.; Carlson, D.A.; Eikum, A.S.; Gjessing, E.T. Electrocoagulation of potable water. *Water Res.* **1984**, *18*, 1355–1360. [CrossRef]
9. Eryuruk, K.; Un, U.T.; Ogutveren, U.B. Electrochemical treatment of wastewaters from poultry slaughtering and processing by using iron electrodes. *J. Clean. Prod.* **2019**, *121*, 1089–1095. [CrossRef]
10. Moreno-Casillas, H.A.; Cocke, D.L.; Gomes, J.A.G.; Morkovsky, P.; Parga, J.R.; Peterson, E. Electrocoagulation mechanism for COD removal. *Sep. Purif. Technol.* **2007**, *56*, 204–211. [CrossRef]
11. Mollah, M.Y.; Morkovsky, P.; Gomes, J.A.; Kesmez, M.; Parga, J.; Cocke, D.L. Fundamentals, present and future perspectives of electrocoagulation. *J. Hazard. Mater.* **2004**, *114*, 199–210. [CrossRef] [PubMed]
12. Johnson, P.N. Ferric chloride and alum as single and dual coagulants. *Acta Sci. Circumst.* **1983**, *75*, 232–239. [CrossRef]
13. Makarov, V.M.; Kalaeva, S.Z.; Zakharova, I.N.; Nevzorov, I.A.; Maltseva, M.S.; Shipilin, A.M.; Krzhizhanovskaya, M.G. Magnetic and X-ray studies of nanodispersed magnetite synthesized from chrome containing galvanic sludge. *J. Nano- Electron. Phys.* **2015**, *7*, 4–12.
14. Wei, X.; Cao, J.; Holub, R.F.; Hopke, P.K.; Zhao, S. TEM study of geogas-transported nanoparticles from the Fankou lead–zinc deposit, Guangdong Province, South China. *J. Geochem. Explor.* **2013**, *128*, 124–135. [CrossRef]
15. Yang, S.; Verdaguer-Casadevall, A.; Arnarson, L.; Silvioli, L.; Čolić, V.; Frydendal, R.; Rossmeisl, J.; Chorkendorff, I.; Stephens, I.E. Toward the decentralized electrochemical production of H_2O_2: a focus on the catalysis. *Acs Catal.* **2018**, *8*, 4064–4081. [CrossRef]
16. Arslan-Alaton, İ.; Kabdaşlı, I.; Vardar, B.; Tünay, O. Electrocoagulation of simulated reactive dyebath effluent with aluminum and stainless steel electrodes. *J. Hazard. Mater.* **2009**, *164*, 1586–1594. [CrossRef]
17. Ghalwa, A.; Nasser, M.; Farhat, N. Removal of abamectin pesticide by electrocoagulation process using stainless steel and iron electrodes. *J. Env. Anal. Chem.* **2015**, *2*, 134. [CrossRef]
18. Garg, K.K.; Prasad, B. Treatment of toxic pollutants of purified terephthalic acid waste water: A review. *Environ. Technol. Innov.* **2017**, *8*, 191–217. [CrossRef]
19. Lu, J.; Li, Y.; Yin, M.; Ma, X.; Lin, S. Removing heavy metal ions with continuous aluminum electrocoagulation: A study on back mixing and utilization rate of electro-generated Al ions. *Chem. Eng. J.* **2015**, *267*, 86–92. [CrossRef]

20. Doğan, D.; Türkdemir, H. Electrochemical oxidation of textile dye indigo. *J. Chem. Technol. Biotechnol.: Int. Res. Process, Environ. Clean Technol.* **2005**, *80*, 916–923.

21. Zongo, I.; Maiga, A.H.; Wéthé, J.; Valentin, G.; Leclerc, J.P.; Paternotte, G.; Lapicque, F. Electrocoagulation for the treatment of textile wastewaters with Al or Fe electrodes: Compared variations of COD levels, turbidity and absorbance. *J. Hazard. Mater.* **2009**, *169*, 70–76. [CrossRef] [PubMed]

22. Holt, P.K.; Barton, G.W.; Mitchell, C.A. The future for electrocoagulation as a localised water treatment technology. *Chemosphere* **2005**, *59*, 355–567. [CrossRef] [PubMed]

23. Ghernaout, D.; Ghernaout, B. On the controversial effect of sodium sulphate as supporting electrolyte on electrocoagulation process: A review. *Desalination Water Treat.* **2011**, *27*, 243–254. [CrossRef]

24. Chen, X.; Chen, G.; Yue, P.L. Separation of pollutants from restaurant wastewater by electrocoagulation. *Sep. Sci. Technol.* **2007**, *42*, 819–833. [CrossRef]

25. Tezcan Un, U.; Topal, S.; Ates, F. Electrocoagulation of tissue paper wastewater and an evaluation of sludge for pyrolysis. *Desalination Water Treat.* **2016**, *57*, 28724–28733. [CrossRef]

26. Eldars, F.M.S.E.; Ibrahim, M.A.; Gabr, A.M.E. Reduction of COD in water-based paint wastewater using three types of activated carbon. *Desalination Water Treat.* **2014**, *52*, 2975–2986. [CrossRef]

27. Zhang, C.; Jiang, Y.; Li, Y.; Hu, Z.; Zhou, L.; Zhou, M. Three-dimensional electrochemical process for wastewater treatment: a general review. *Chem. Eng. J.* **2013**, *228*, 455–467. [CrossRef]

28. Jung, K.-W.; Hwang, M.-J.; Park, D.-S.; Ahn, K.-H. Performance evaluation and optimization of a fluidized three-dimensional electrode reactor combining pre-exposed granular activated carbon as a moving particle electrode for greywater treatment. *Sep. Purif. Technol.* **2015**, *156*, 414–423. [CrossRef]

29. Xiong, Y.; Strunk, P.J.; Xia, H.; Zhu, X.; Karlsson, H.T. Treatment of dye wastewater containing acid orange II using a cell with three-phase three-dimensional electrode. *Water Res.* **2001**, *35*, 4226–4230. [CrossRef]

 *water*

Article

Textile Wastewater Purification Using an Elaborated Biosorbent Hybrid Material (*Luffa–Cylindrica*–Zinc Oxide) Assisted by Alternating Current

Amina Othmani [1,2]**, Aida Kesraoui** [1]**, Roberto Boada** [3]**, Mongi Seffen** [1,*] **and Manuel Valiente** [3,*]

[1] Laboratory of Energy and Materials (LabEM): LR11ES34, Higher School of Science and Technology of Hammam Sousse, University of Sousse, 4011 Hammam Sousse, Tunisia

[2] Faculty of Sciences of Monastir (Monastir University), 5000 Monastir, Tunisia

[3] GTS-UAB Research Group, Department of Chemistry, Faculty of Science, Universitat Autònoma Barcelona, 08193 Bellaterra, Spain

* Correspondence: mongiseffen@yahoo.fr (M.S.); Manuel.Valiente@uab.cat (M.V.); Tel.: +216-26554812 (M.S.)

Received: 31 May 2019; Accepted: 22 June 2019; Published: 27 June 2019

Abstract: This paper aims to synthesize hybrid materials with high pollutant-uptake capacity and low costbased based on *Luffa cylindrica* (*L.C*) and different percentage of Zn^{2+} in the presence and absence of alternating current (AC). Physico-chemical, morphological and structural characterizations of the hybrid materials were performed by Boehm method, point zero charge (pH_{pzc}), infrared characterizations (IR), scanning electron microscopy (SEM), energy–dispersive spectroscopyand and X-ray photoelectron spectroscopy. The efficiency of the designed hybrid materials was optimized based on their performance in water depollution. Methylene blue (MB) and industrial textile wastewater were the investigated pollutants models. IR characterizations confirmed the fixation of Zn^{2+} onto the *L.C* by the creation of Zn-OH, Zn-O and Zn-O-C bonds. Boehm titration showed that the fixation of Zn^{2+} onto *L.C* is accompanied by an increase of the basic functions of its surface and subsequently an increase in the pH_{pzc}. SEM results confirmed the fixation of Zn^{2+} onto the *L.C* coupling AC with biosorption showed an increase in the adsorbed amount of MB and speed when adding the 4% of Zn^{2+} compared to the pure *L.C* the Q_m shifted from 3.22 to 9.84 and 8.81 mg/g, respectively, for hybrid materials synthesized under AC, in absence of AC and pure *L.C*.

Keywords: alternating current; coupling; hybrid material; biosorption; wastewater reuse

1. Introduction

Textile dying wastewaters are classified among the most highly toxic effluents [1,2]. Numerous processes have been suggested for treatment and purification such as biological treatment [3], coagulation-flocculation [4], adsorption [5], ultrafiltration [6], electrocoagulation [7], reverse osmosis [8] and anodic oxidation [9]. The current research was devoted to water remediation through the valorization of the abundant renewable resource of cellulosic fibers [10]. Brown algae [11], *Bacillus maceran* [12], *Posidonia oceanica* [13], corn stigmas [14], *Agave Americana* [15], *Luffa cylindrica* [16], fly ash and red mud [17], raw date pits [18] and *Phragmites australis* [19] are widely used as a natural and cheap biosorbent for a pollutant removal. However, achieving a quick adsorption kinetic with a high possibility of water reuse after filtration absolutely depends on the biosorbent efficiency.

Hybrid materials are praised for not only intermediate properties between mineral and organic matter but also for new interesting behaviors that allow them to be used in several fields of applications. Such domains include catalytic applications, medical and pharmaceutical applications, optoelectronics, the environment, and biomaterials [20,21]. Hybrid materials can be prepared by several methods such as chemical vapor deposition (CVD) [22], physical vapor deposition (PVD) [23], laser ablation and sputtering [24]. The synthesis of hybrid materials based on cellulose and zinc oxide is the objective of recent studies. Perelshtein et al. (2009) developed a composite based on cotton fibers and zinc oxide. They synthesized the ZnO particles and then deposited them on the cotton surface by ultrasonic irradiation (sonochemical method) [25]. Weili et al. (2010) have succeeded in synthesizing a hybrid matrix based on wurtzite-cellulose ZnO bacteria by the thermal decomposition method. This new hybrid material obtained plays a very important role in improving the photo catalytic activity of anionic dyes such as, orange methyl [26].

However, the techniques used for the synthesis of hybrid materials are expensive. Other researchers have developed less expensive techniques such as the electrochemical method [27], spray pyrolysis [28] and sol-gel process [29]. Recently, Kesraoui et al. (2018) reported that the precipitation method can be a simple and cheap method used for the synthesis of hybrid materials. They have successfully synthesized hybrid material base on *L.C* and metal oxides (ZnO, Al_2O_3). These last showed a good ability for dyes removals compared to the pure *Luffa cylidnrica* (*L.C*) [30]. According to literature, the synthesis of hybrid materials based on compounds containing cellulose and ZnO can present interesting properties allows it to be used in several applications.

However, the removal process often requires a lot of time to reach the balance. Therefore, coupling environmental security (there is no transformation of the initial molecule to toxic compounds as for the oxidation process), low cost, final efficiency, and quick process remains a very important stake.

Taking into account these considerations, a set of aims are proposed in this paper, the synthesis of a high performing hybrid material based on *Luffa cylinrdica* (*L.C*) and metal oxide with high pollutant-uptake capacity and low cost by an easy precipitation method in absence and presence of alternating current (AC). Since their good affinity for the removal of several pollutants and their interesting physical and chemical properties, *L.C* and ZnO were considered for the synthesis of the hybrid materials [31,32]. Two main pollutants were taken as models in this study; a cationic dye the methylene blue (MB) taken as a model of dye and an industrial textile wastewater. A detailed investigation of the physicochemical, morphological and structural characterizations was carried out using Boehm titration. Also, Drift method, scanning electron microscopy (SEM), Energy-dispersive spectroscopy (EDS), infrared spectroscopy (IR) and X-ray photoelectron spectroscopy (XPS) analysis were performed. The possible pathway of the Zn^{2+} fixation into the *L.C* fibers and the mechanism of possible interaction of MB onto the lignocellulosic surface are suggested. The mathematic modeling was also supplied using the stochastic model of Brouers-Sotolongo.

2. Material and Methods

2.1. Materials

Methylene blue (MB) (purity: 95%) from Fluka™ (manufacturer, Streinheim, Germany), with chemical formula $C_{16}H_{18}N_3SCl$ and molar mass 319.85 g/mol, was chosen as commercial dye model for this study.

Zinc nitrate ($Zn(NO_3)_2.6H_2O$; Hexahydrate Purified), sodium chloride (NaCl), sodium bicarbonate ($NaHCO_3$), sodium carbonate (Na_2CO_3) and sodium hydroxide (NaOH) (purity 99%) were obtained from LOBA CHEMIE (Wodehouse road, Mumbai, India).

The biosorbent used for the synthesis of the hybrid material was fruits of Tunisian *Luffa cylindrica* fibers. This biosorbent was purchased at the local market Sousse. The chemical composition of these fibers was determined by Kesraoui et al. (2016). These fibers are composed of 54% of cellulose, 11% of lignin, 5% of pectin, 7% of fats and waxes and 23% hemicelluloses [33].

2.2. Methods

2.2.1. Biosorbent and Adsorbate Preparation

Tunisian *Luffa cylidnrica* (*L.C*) was chosen as a natural biosorbent in this paper. The preparation of this biosorbent consisted of cutting the fibers finely, washing them several times to remove all impurities and drying them at 70 °C until the material was completely dried. As for the adsorbate preparation, it consisted of dissolving 10 mg of MB in 1 L of distilled water to obtain the desired concentration (10 mg/L).

2.2.2. Preparation of Hybrid Materials *L.C*+ (1%, 2%, and 4% Zn^{2+}) in Presence and Absence of AC

The Zn^{2+} precursors of solution were obtained from ($Zn(NO_3)_2 \cdot 6H_2O$ (Hexahydrate Purified) Sigma-Altrich (99%) (Wodehouse road, Mumbai, India).

The preparation of the hybrid material with different percentages of Zn^{2+} (1%, 2%, and 4%) consists inmixing 5 g of *L.C* (size 250 μm) with 0.05, 0.1 and 0.2 g of $Zn(NO_3)_2 \cdot 6H_2O$ respectively. Each composite was dissolved with the biomass in 100 mL of distilled water at 298 K at pH = 10, under stirring for two hours. The size of the fibers used was chosen after sieving using an electric sifter and after studying the effect of fiber grain size (40, 80, 125 and 250 μm on the biosorption efficiency). Then, the product was washed several times using distilled water to remove the excess of $Zn(NO_3)_2 \cdot 6H_2O$ which has not been fixed onto the surface of *L.C*. Thereafter, the product obtained was transferred to a sand bath to dry it at 393 K for two hours. A similar experiment was done for the synthesis of the hybrid material under AC using two zinc electrodes (1.3 × 2.5 cm^2; purity 99%) immersed into the solution while stirring. The electrical mounting comprising an AC source, a voltmeter to fix the current density at 0.5 A/m^2 and the voltage at 15 volts.

2.3. Morphological and Crystallographic Characterizations of Hybrid Materials

Morphological and structural characterizations were done in order to characterize the hybrid material synthesized in the presence and absence of AC.

Infrared characterizations (IR) were performed using a Perkin Elmer Spectrum using KBr pellet technique in the frequency range of 4000 to 500 cm^{-1}. Scanning electron microscopy (SEM) was performed using a JEOL JSM 5400 scanning microscope (USA) after coating them with gold using a JEOL JFC-1199E ion sputtering device (USA). Energy dispersive spectroscopy (EDS) was planned to assess the surface elemental compositions of raw and the hybrid materials *L.C* +4% Zn^{2+} using a JEOL JSM 5400 scanning microscope. EDS was performed after coating them with gold using a JEOL JFC-1199E ion sputtering device. X-photoelectron spectroscocopy (XPS) measurements were performed at room temperature with a SPECS PHOIBOS 150 hemispherical analyzer (SPECS GmbH, Berlin, Germany) in a base pressure of 5 × 10^{-10} mbar using monochromatic Al Kalpha radiation (1486.74 eV) as an excitation source.

2.4. Quantitative and Qualitative Characterization of the Pure L.C and the Hybrid Material L.C-Zn^{2+}

The physicochemical characterization of the pure *L.C* and the synthesized hybrid materials was determined by the Drift method and the Boehm titration. The determination of the zero charge point pH (pH_{pzc}) consisted of placing 0.1 g of *L.C* in 30 mL of NaCl solution (0.01 M) at different pH ranging from 2 to 12. The initial pH is obtained by adding a certain amount of NaOH or HCl (1 M). For the first 24 h, the pH must be measured and then this operation must be repeated during the second 24 h in order to register the difference in pH. By plotting the pH_f = f (pH_i) curve, the pH_{pzc} corresponds to the intersection of this curve with the straight line pH_f = pH_i [34].

The determination of acid-basic properties by Boehm method consisted in bringing onto contact 0.5 g of each lignocellulosic material with 25 mL of one of these bases: $NaHCO_3$, Na_2CO_3 and NaOH

(0.1 M) for 48 h with stirring. 10 mL of each solution was back-dosed with NaOH solution (0.1 M) after acidification with an excess of 0.1 M of hydrochloric acid [35].

2.5. Biosorption of MB onto the Synthesized Hybrid Material

The experiments used for the MB removal by biosorption have been performed in batch reactor by adding 0.1 g of adsorbent (Pure *L.C*, hybrid material elaborated by precipitation) in 100 mL of MB solution (pH = 10, C_i = 10 mg/L, J = 0.5A/m^2, voltage = 15 volts, T = 298K). The electrical mounting comprising an AC source and a voltmeter. All the analyses were carried out in triplicate.

After biosorption, residual concentrations were determined by UV visible spectrophotometer biochrome Libra S.22 at λ = 663 nm based on the following equation [36]. Plastic and quartz cuvettes were used for absorbance tests.

$$\text{Dye removal (\%)} = \frac{(C_i - C_t)}{C_i} \times 100 \tag{1}$$

where C_i is the initial dye concentration (mg/L) and C_t is the dye concentration at any time (mg/L).

2.6. Kinetics Studies

The evaluation of the MB biosorption in terms of adsorbed quantity was done following the Equation (2) [37].

$$Q_t = \frac{C_i - C_e}{m} \times V \text{ (mg/g)} \tag{2}$$

where Q_t is the adsorbed quantity at equilibrium time, C_i is the initial MB concentration (mg/L), C_e is the residual MB concentration at any time (mg/L), V is the volume of solution (L) and m is the mass of the adsorbent (g). At equilibrium, C_i is equal to C_e and Q is equal to Q_e [38].

Brouers-Sotolongo (B.S) model was used to fit the experimental data of MB biosorption. The generalized kinetic equation of the Brouers-Sotolongo model has been developed to provide a universal function for the kinetics of complex systems characterized by exponential law and/or power exponent behaviors. This kinetic model unifies and generalizes previous theoretical attempts to describe what is called "fractal kinetics". The mathematical development of this model and its application are presented in Brouers and Sotolongo [39–41].

The pseudo BSf (n, α) sorption kinetics is given by Equation (3):

$$Q_t = Q_e[1 - (1 + (n-1)\left(\frac{t}{\tau_c}\right)^{\alpha})^{\frac{-1}{n-1}}] \tag{3}$$

where, n is order of the fractional reaction, α is "fractal time" exponent.

τ_c: characteristic time, Q_e: the adsorbed quantity at saturation, Q_t: the adsorbed quantity at any time.

2.7. Evaluation of the Treated Water Qualities

2.7.1. Determination of Chemical Oxygen Demand (COD) and Total Organic Carbon (TOC)

The global mineralization was determined by measuring the chemical oxygen demand (COD) by a photometric method. The % COD removal can be estimated by (Equation (4)) [37].

$$\text{COD removal (\%)} = \frac{COD_i - COD_t}{COD_i} \times 100 \tag{4}$$

where COD_i is the initial COD (g/L O_2) and COD_t is the residual COD at any time (g/L O_2).

The total organic carbon (TOC) in the solutions of MB and industrial wastewater was measured by a TOC analyzer (HACHIL, 550-TOC-TN model, Schwerte, Germany) and the TOC removal can be estimated as follows:

$$\text{TOC removal (\%)} = \frac{\text{TOC}_i - \text{TOC}_t}{\text{TOC}_i} \times 100 \tag{5}$$

where, TOC_i and TOC_t are respectively the initial total organic carbon and the total organic carbon obtained after fixed time t of electrolysis treatment (ppm C.O).

2.7.2. Germination Tests

The phytotoxicity test consists in the determination of the inhibitory effect of the treated water on the germination and growth potential of lettuce based on the GI. All sample experiments, including reference (pure water), were tripled. The results obtained are finally expressed according to the following relation [42]:

$$\text{GI(\%)} = \frac{\text{number of seeds sprouted in the sample}}{\text{number of seeds sprouted in the reference}} \times \frac{\text{averagelength of root in the sample}}{\text{average root length in the reference}} \times 100 \tag{6}$$

3. Results and Discussion

3.1. Surface Characterizations

3.1.1. Energy-Dispersive Spectroscopy (EDS) Analysis

The zinc deposit was analyzed with EDS (Table 1). Carbon and oxygen were the major compounds of *L.C* and hybrid materials due to their chemical composition (cellulose, hemicellulose, and lignin). In addition, the appearance of the Zn atom confirmed its fixation onto the surface of *L.C*. The AC enhanced the distribution of Zn^{2+} into the *L.C* fibers where the rates according to the mass of Zn reached about 1.19% and 0.95% respectively for the hybrid material synthesized in presence and absence of AC.

Table 1. Percentages by mass of the chemical elements present on the fiber surface of the raw *Luffa cylindrica* (*L.C*) and the hybrid materials *L.C* + 4% Zn^{2+} and *L.C* + 4% Zn^{2+} + AC.

	C	O	Na	Mg	Al	Si	P	S	Cl	Ca	Mo	Zn
L.C	54.46	40.60	0.45	0.38	0.80	1.10	0.29	0.00	0.35	1.40	0.17	0.00
L.C + 4% Zn^{2+} + AC	57.01	35.64	0.00	0.38	4.78	0.00	0.00	0.00	1.00	0.00	0.00	1.19
L.C + 4% Zn^{2+}	50.43	40.18	0.00	0.79	4.58	2.40	0.00	0.08	0.59	0.00	0.00	0.95

3.1.2. Infrared Spectroscopy (IR)

Figure 1 shows that the spectrum of pure *L.C* has a vibration band at 3438 cm^{-1} which corresponds to the O-H bond present in cellulose, hemicellulose, and lignin. In the presence of different percentages of Zn^{2+} (1%, 2%, and 4%), obtained by precipitation (Figure 1a), the band of hydroxyl groups is observed at 3435, 3430 and 3414 cm^{-1}. The decrease in the position of these bands can be attributed to the presence of elongation vibration of the Zn-OH bond, between Zn^{2+} and cellulose [43]. These results indicate that the hydroxyl groups have a strong interaction with Zn^{2+}.

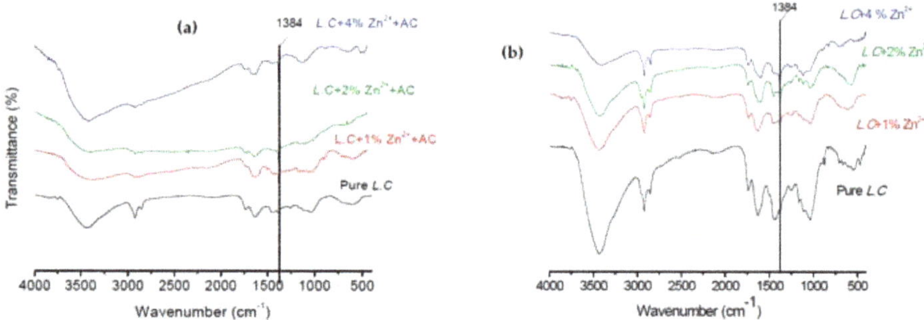

Figure 1. Spectrum of pure *L.C* and *L.C* + (1%, 2% and 4% Zn^{2+}) synthesized (**a**) in presence of alternating current (AC) and (**b**) in the absence of AC.

The appearance of 1383 peaks at 4% *L.C*-Zn^{2+} is probably due to symmetric Zn-O-C vibration [44]. The bands at 1032.56, 1016.77 and 1016 cm^{-1}, which correspond to the vibration of the C-O bond, increase in intensity, which can be attributed to the appearance of the Zn-O-C bond.

Based on Figure 1b, the rising of the percentages of Zn^{2+} added during the synthesis of hybrid materials by precipitation under AC caused the decrease of the position of the band of hydroxyl groups. This last passed from 3438 cm^{-1} to 3430, 3427and 3420 cm^{-1}. This decrease in position can be attributed to the formation of Zn-OH stretching vibration [45].

Furthermore, we note the appearance of the Zn-O bond is proven by the appearance of peaks at 1383, 517 and 485 cm^{-1} [46].

The bonds located at 1032.56, 1016.77 and 1016 cm^{-1} correspond to the vibration of the C-O bond. An increase in the intensity of these bonds was shown when adding the Zn^{2+}. This increase can be explained by the appearance of the Zn-O-C bond.

The band located at 1383 cm^{-1} presents the main difference between the IR spectrum obtained the hybrid materials synthesized in the presence and absence of AC. For the hybrid material synthesized under AC, the band at 1383 cm^{-1} appears from 1% of Zn^{2+}. While the hybrid material synthesized by precipitation this band appears only for 4 % of Zn^{2+}. This behavior is probably due to the effect of AC on the fixation of Zn^{2+} ions into the *L.C* fibers by the creation of Zn-O-C bonds.

3.1.3. X-ray Photoelectron Spectroscopy (XPS) Analysis

XPS measurements were performed in order to clarify the main difference between the syntheses of the hybrid material with and without AC. Based on IR characterizations; Zn is strongly bound to the lignocellulosic surface by OH hydroxyl ions which are confirmed by the presence of O-Zn-O and O-Zn bonds. Figure 2 displays the main XPS results for the hybrid material *L.C* + 4% Zn^{2+} synthesized in the presence and absence of AC. The high-resolution XPS spectra of C(1s), O(1s), Zn(2p), Zn LMM^{+} show noticeable differences between the material synthesis by the two methods. However, neither of them showed any impurities. The C(1s) peak, depicts a chemical shift by +/−0.5 eV in comparison with the C(1s) spectra observed for the hybrid material synthesized in absence of AC (not AC) (282.5 eV) towards a relatively high binding energy positioned at 283 eV specific to C=O and O–C–O [47]. Similar behavior was observed for the O(1s) peak which is positioned at the binding energies of 529.9 and 531.2 eV for the hybrid material *L.C* + 4% Zn synthesized in absence and presence of AC, respectively. These contributions can be assigned to O^{2-} ions in the Zn-O bonding and to O–H groups of adsorbed water molecules [48]. The four peaks observed at the binding energies of 1017.5, 1020 and 1040 and 1044.5 eV, correspond respectively to the zinc oxide (ZnO) and the spin orbit of Zn (2p$_{3/2}$) and Zn (2p$_{1/2}$) for the material synthesized in absence of AC [48]. In presence of AC, two main peaks located at 1021 and 1043.5 eV are assigned to the Zn(2p$_{3/2}$) and Zn(2p$_{1/2}$) energy levels.

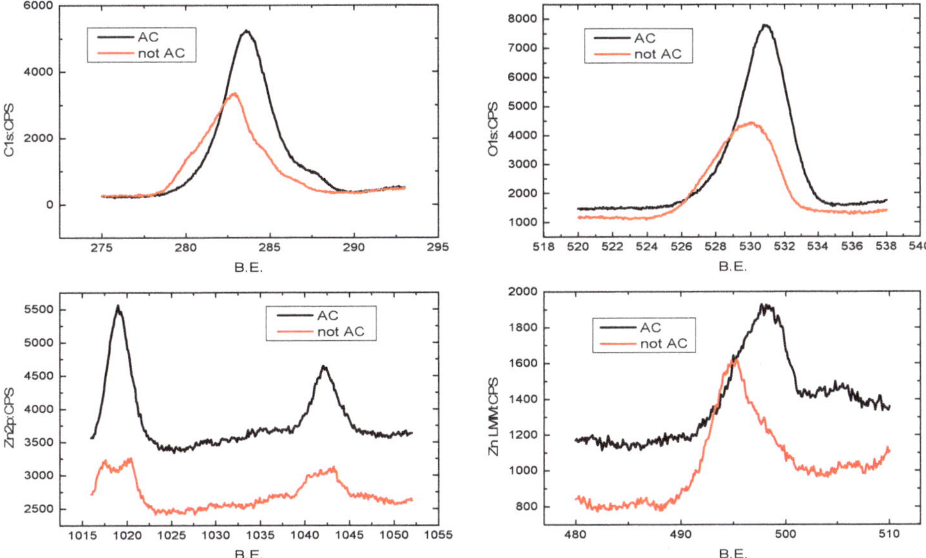

Figure 2. Spectra of C1S, O 1s, Zn 2p and Zn LMM of various morphologies of *L.C* + 4% Zn^{2+} synthesized in presence of AC (-) and absence of AC (not AC -).

Both of these were symmetrical and narrow, indicating the absence of the zinc oxide. Despite this difference, these peaks of the Zn(2p) detected in both hybrid materials are related to Zn-O bonding as confirmed by previous studies [49]. Furthermore, we determined a small offset followed by a big discrepancy in terms of the intensities of peaks which were higher in case of the use of AC. These findings indicated that the obtained Zn does not occur in the same oxidation state and even if they are similar in number. It can also be attributed to the depth distribution of the atoms. This increase in intensity is explained by the diffusion of atoms towards the surface which was more pronounced compared to the hybrid material synthesized without AC. This behavior might stem from the difference in the local chemical environment. It can also be directly related to the number of atoms in the corresponding chemical state, which is confirmed by the signal Zn LMM Auger region. Another possibility is the sharing of the element with its neighbor atoms which applied a crystal field on it. This can also be interpreted in terms of the asymmetry in the bond arrangement which can create extra pressure on the element—so, the binding energy required for the ejection of the electron is slightly higher than its previous case. In the light of these results, we confirm the fixation of Zn^{2+} onto the *L.C* by the creation of Zn-O bond as proven by IR characterizations for both methods used for the synthesis of the hybrid material. However, the environment of Zn bonding is slightly different.

3.1.4. Scanning Electron Microscopy (SEM) Analysis

SEM characterizations were performed to gain further insight into the distribution of Zn^{2+} obtained from $(Zn(NO_3)_2 \cdot 6H_2O)$ on the *L.C* surface by precipitation in presence and absence of AC. Figure 3 shows the obtained micrograph of the pure *L.C* (Figure 3a), hybrid material (*L.C* + 4% Zn^{2+}) synthesized by precipitation (Figure 3b), and hybrid material (*L.C* + 4% Zn^{2+}) synthesized under AC (Figure 3c). *L.C* presents a rough structure and a homogeneous appearance which is formed by bonded multicellular fibers (Figure 3a). This indicates that the existence of a large number of hydroxyl groups could provide effective interaction between *L.C* and Zn^{2+} [26,50]. Figure 3b,c refer to the hybrid material (*L.C* + 4% Zn^{2+}) synthesized in the absence and presence of AC. Here the Zn^{2+} deposit is well visible and well-fixed onto the *L.C* and exhibits good dispersion without significant aggregation. The use of AC has enhanced the distribution of Zn^{2+} into the *L.C* fibers.

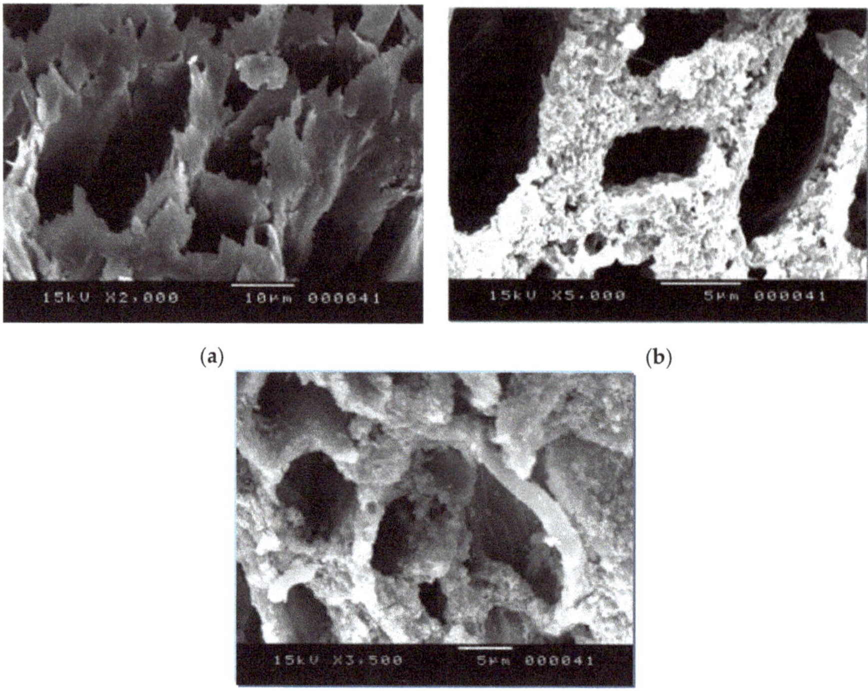

(a) (b)

(c)

Figure 3. Images of (**a**) pure *L.C*, (**b**) *L.C* + 4% Zn^{2+} synthesized in absence of AC, (**c**) *L.C* + 4% Zn^{2+} synthesized in presence of AC.

3.1.5. Surface Chemical Analysis of Pure *L.C* and Hybrid Materials

Table 2 shows the main pH_{pzc} values obtained for the pure *L.C* and the synthesized hybrid materials. The presence of zinc increased the pH_{pzc} which made the fibers slightly basic. In addition, pH_{pzc} increases with increasing the percentage of zinc added.

Table 2. pH_{pzc} of hybrid material synthesized in absence and presence of AC.

Samples	pH $_{pzc}$ under AC	pH $_{pzc}$ without AC
Pure *L.C*	7.38 ± 0.021	7.38 ± 0.021
L.C + 1% Zn^{2+}	7.57 ± 0.008	7.41 ± 0.016
L.C + 2% Zn^{2+}	7.77 ± 0.005	7.69 ± 0.012
L.C + 4% Zn^{2+}	7.96 ± 0.005	7.86 ± 0.012

The pH_{PZC} of *L.C* is equal to 7.38. This result shows that the surface of *L.C* has a positive charge at pH values below pH_{PZC} and should, therefore, be able to adsorb anions and a negative charge at pH values above pH_{PZC} and should, therefore, be able to adsorb cations [51].

On the other hand, according to Table 3, for the pure *L.C*, the amount of base and the total amount of acid groups are approximately equal. These results suggest that *L.C* fibers have an amphoteric character. This amorphous character confirms the result found by the pH_{PZC} (7.38).

Table 3. Different functional groups obtained on the adsorbent surfaces.

Adsorbent	Acidic Function ($\times 10^{-4}$ mmol/g)				Basic Function ($\times 10^{-4}$ mmol/g)
	Carboxylic	Phenolic	Lactonic	Total	
Pure *L.C*	0.397 ± 0.002	0.935 ± 0.003	0.178 ± 0.008	1.510 ± 0.005	1.560 ± 0.0008
L.C + 1%Zn^{2+}	0.280 ± 0.002	0.790 ± 0.004	0.200 ± 0.001	1.280 ± 0.001	1.610 ± 0.0005
L.C + 2% Zn^{2+}	0.201 ± 0.002	0.544 ± 0.003	0.310 ± 0.002	1.055 ± 0.002	1.700 ± 0.0004
L.C + 4% Zn^{2+}	0.170 ± 0.002	0.400 ± 0.002	0.430 ± 0.002	1.000 ± 0.001	1.810 ± 0.0007
L.C + 1% Zn^{2+} + AC	0.220 ± 0.001	0.503 ± 0.001	0.270 ± 0.002	0.993 ± 0.0009	1.560 ± 0.0006
L.C + 2%Zn^{2+} + AC	0.198 ± 0.0008	0.399 ± 0.002	0.390 ± 0.0008	0.987 ± 0.0006	1.790 ± 0.0003
L.C + 4%Zn^{2+} + AC	0.155 ± 0.001	0.236 ± 0.001	0.511 ± 0.0005	0.902 ± 0.0005	1.920 ± 0.0001

The carboxylic and phenolic functions of hybrid materials synthesized under AC (*L.C* + Zn^{2+} + AC) are lower than those synthesized in absence of AC (*L.C* + Zn^{2+}) (0.155 ± 0.001 and 0.170 ± 0.002 for the carboxylic functions, 0.400 and 0.230 for the phenolic functions, respectively). These findings confirm the increase of pH$_{pzc}$ obtained during the synthesis of the hybrid material in presence of AC (7.96 ± 0.005) which was slightly higher than that obtained during the synthesis in its absence (7.86 ± 0.012). It is apparent that the addition of Zn^{2+} and the use of AC slightly influenced the functional groups of hybrid materials obtained. Such behavior can be explained by the increase in the number of available sites on the surface of the materials, which enhances the uptake capacity of such materials and accelerates the biosorption kinetics.

Furthermore, the basic and lactonic functions of *L.C*-Zn^{2+} + AC are higher than those of *L.C*-Zn^{2+} (1.920 ± 0.0001 and 1.810 ± 0.0007 for basic functions and 0.511 ± 0.0005 and 0.430 ± 0.002 for lactonic functions, respectively). According to Mahdoudi et al. (2015), the increase in basic functions is probably due to partial or total deprotonation of the active sites of the hybrid material. Therefore, the fixation of zinc onto *L.C* is accompanied by an increase of the basic functions of its surface and subsequently an increase in the pH$_{pzc}$. The dissociation of the surface oxygen groups of the acidic groups (carboxylic, lactone and phenol) confers a negative charge on the surface of Luffa cylindrica. Thus, the surface acid sites are of Brönsted type [52]. With regard to the positive charge, this may result from the existence of basic oxygen groups such as pyrones or benzopyran.

3.2. Kinetics of MB Bbiosorption Assisted by AC

The MB biosorption tests were performed at an initial dye concentration of 10 mg /L, pH = 10, a mass of the adsorbent of 0.1 g and a temperature of 298 K, V = 15 volts, J = 0.5 A/m^2.

Figure 1 shows that the method of elaboration of the hybrid material has a great influence on the biosorption capacity. Indeed, the adsorbed quantity of the highest MB is obtained by the hybrid material (*L.C* + 4%Zn^{2+} + AC). As a result, the adsorbed amount of MB increased from 3.22 mg/g for pure *L.C* to 8.81 mg/g for *L.C* + 4% Zn^{2+} and 9.84 mg/g for *L.C* + 4% Zn^{2+} + AC. Similarly, the MB removal rates reached about 98.4, 88.1 and 31.4% respectively for the hybrid material *L.C* + 4%Zn^{2+} synthesized in the presence and absence of AC and the pure *L.C* (Figure 4).

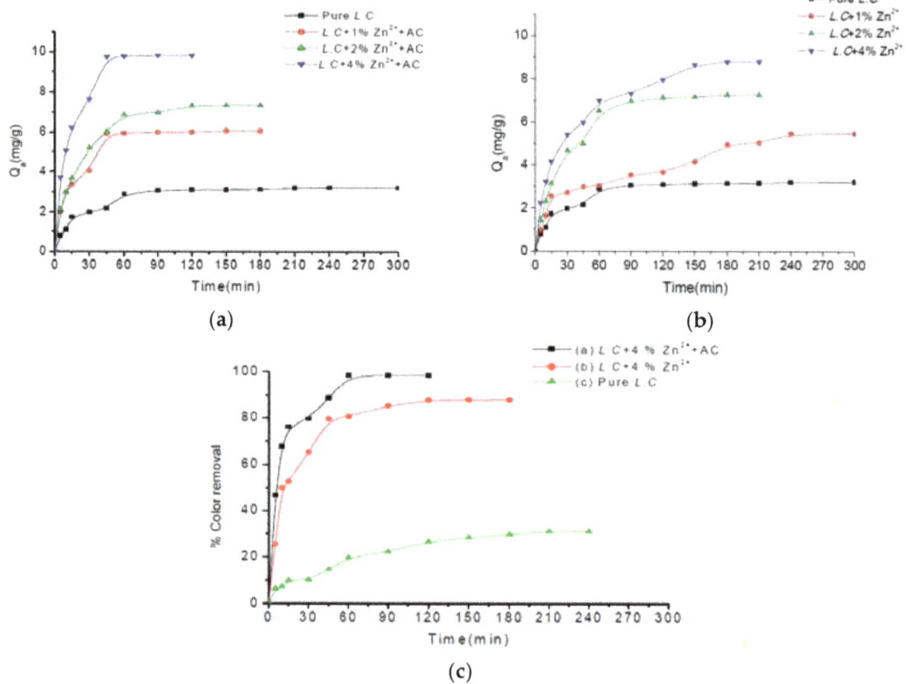

Figure 4. Kinetics of retention of methylene blue (MB) onto different materials used *L.C*; pH = 10, C_i = 10 mg/L, m = 0.1 g, T = 298 K. (**a**) hybrid material *L.C* + 4% Zn^{2+} synthesized in presence of AC, (**b**) hybrid material *L.C* + 4% Zn^{2+} synthesized in absence of AC, (**c**) % color removal using different materials.

The use of AC in the synthesis of the hybrid material for different percentages of Zn^{2+} improved the quantities adsorbed as well as the time required for the process. This behavior can be explained by the facility of the access of pollutants into the biosorbent pores and the availability of more active sites for biosorption offered by the synthesized hybrid materials.

Contrariwise, the best results were obtained for hybrid materials *L.C* + 4% Zn^{2+} synthesized in the presence and absence of AC as the Q_{max} reached, respectively, 9.83 mg/g and 8.81 mg/g. Interestingly, the high amount of added Zn^{2+} assures the increase of superficial negatively charged groups obtained by lactonic function (as confirmed by Boehm titration). Therefore, this increase of negatively charged groups increased the biosorption capacity.

When comparing biosorption when using hybrid material synthesized in the presence of AC and those synthesized in its absence as well as the pure *L.C*, definitely, the speed of biosorption and the capacity uptake were greatly enhanced. While the time required for the biosorption of MB is almost lower than that required when using the pure *L.C* and hybrid materials synthesized in the absence of AC. This is probably due to the existence of the electric field which increased the speed of movement of the molecules of MB and facilitated their access into the active sites.

3.3. UV Characterizations

To forecast the main phenomena that may occur during the MB biosorption, an UV-visible characterization was performed as shown in Figure 5. The characterization was done in order to compare the initially existing bands and the new ones that may appear during the biosorption of MB assisted by AC. The initial MB spectra display three main peaks located at 293,602 and

663 nm corresponding to the chromophore (dimethylamino group) and the aromatic rings in the MB molecule [53]. After MB biosorption, a decrease in the intensity of peaks recorded in the visible range at λ_{max} = 663 nm relative to the naphthalene group [54] was observed for all biosorbent used. No new band appeared, thus indicating that no intermediates appeared and confirming that the color removal is probably due to a biosorption and not to an oxidation process which was probably due to the azo groups. It is interesting to note that the biosorption of MB was particularly important by using the hybrid material *L.C* + 4% Zn^{2+} synthesized by precipitation under AC (Figure 5a) compared to the pure *L.C* (Figure 5c) and hybrid material synthesized in absence of AC (Figure 5b). In addition, the biosorption was rapidly obtained and within 120 min and a steady state concentration was achieved indicating a great ability of hybrid material for dye removal. On the contrary, the MB biosorption was tardier and the removal efficiency was lower when using the pure *L.C* (Figure 5c).

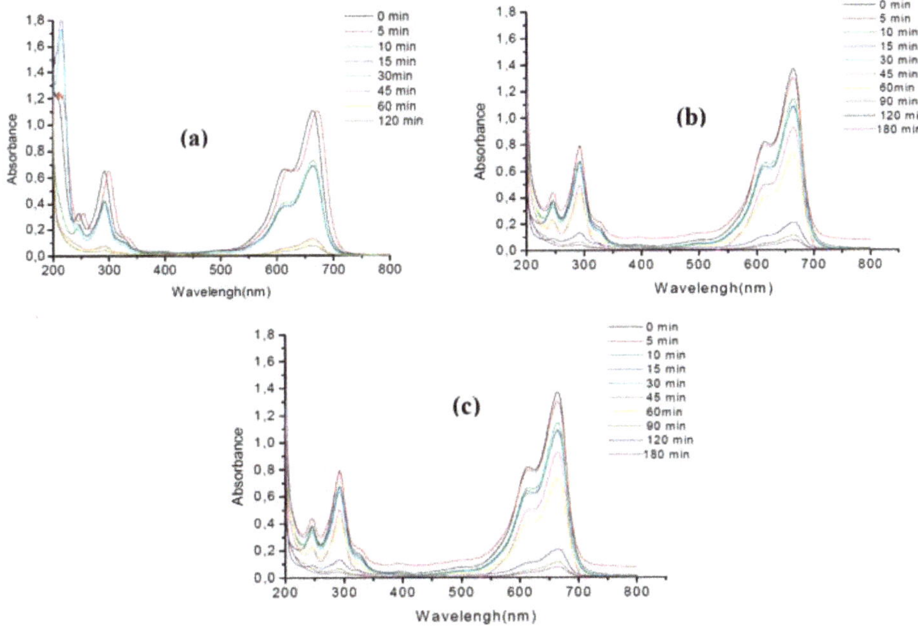

Figure 5. Spectra of MB biosorpion assisted by AC using (**a**) hybrid material *L.C* + 4% Zn^{2+} synthesized in presence of AC, (**b**) hybrid material *L.C* + 4% Zn^{2+} synthesized in absence of AC and (**c**) pure *L.C*, C_i = 10 mg/L, pH = 10, $m_{L.C}$ = 0.1g, V = 0.1 L, T = 298 K.

3.4. Modeling of the Biosorption of MB Assisted by AC Using Brouers Sotolongo (B.S) Model

The biosorption mechanism of MB onto pure *L.C* and hybrid materials elaborated by precipitation in the presence and absence of AC was studied by B.S model. The surface of the biosorbent is heterogeneous. The functions are distributed randomly hence the need to use a stochastic model (B.S). This model was chosen based on the coefficient of determination (R^2), the non-linear chi-square test (χ^2) and the calculated Q_e (Table 4 and Table S1 in the complementary information).

Table 4. Comparative results for the B.S model.

Biosorbent Used	Q_{exp} (mg/g)	Q_{the} (mg/g)	R^2	χ^2	α	τ_c
Brouers Sotolongo n = 1						
Pure *L.C*	3.22	4.33	0.9999	0.0001	0.99	91.52
L.C + 1% Zn^{2+} + AC	6.06	5.97	0.9995	0.0005	1.11	56.14
L.C + 2% Zn^{2+} + AC	7.33	7.41	0.9999	0.0005	1.27	47.98
L.C + 4% Zn^{2+} + AC	9.84	9.31	0.9999	0.0001	5.41	36.83
L.C + 1% Zn^{2+}	5.47	5.40	0.9999	0.0001	0.99	59.01
L.C + 2% Zn^{2+}	7.27	7.22	0.9999	0.0004	1.06	49.13
L.C + 4% Zn^{2+}	8.81	8.64	0.9999	0.0001	3.64	39.55

Table 4 shows that B.S model BSf (1, α) presents the best fit to experimental data related to MB biosorption by *L.C* and hybrid materials. Indeed, the B.S model BSf (1, α) has the highest coefficient of determination and the lowest nonlinear chi-square test. In addition, the Qe calculated by this model gives acceptable values.

The comparison of the values of τ_c calculated from BSf (1, α) shows that *L.C* + 4% Zn^{2+} + AC has the lowest value of τ_c (36.83). These results indicate that the AC synthesized hybrid materials remarkably improve both the adsorption capacity and the reaction rate of pure *L.C*.

On the other hand, the fractal constant α of *L.C* is less than one—which shows that the surface of *L.C* is heterogeneous [55]. Although the fractal constant α of the hybrid materials exceeds one, the kinetic is not clearly fractal [56].

For the hybrid materials, the fractal constant α increases with the increase of the percentage of Zn^{2+} added. The hybrid material elaborated by AC (*L.C* + 4% Zn^{2+}) presents a higher value of α. In fact, Selmi et al. (2018) showed that the fractal constant α increases with the number of functional surface groups [57]. Therefore, the hybrid material elaborated by AC (*L.C* + 4% Zn^{2+}) has the highest number of functional groups.

3.5. Evaluation of the Elaborated Hybrid Material for MB Biosorption: A Comparative Study

To estimate the efficiency of the hybrid material synthesized by the precipitation method in presence and absence of AC and other low-cost material widely used for dye removal, a comparative study in terms of biosorption capacity (Q_m) and efficiency was done as illustrated in Table 5. Results obtained when using the hybrid material *L.C* + (1%, 2%, and 4% Zn^{2+}) synthesized by precipitation under AC are largely higher than those obtained when using the hybrid material synthesized in absence of AC and the other low-cost materials. The Q_m shifted from 3.22 to 8.81 and 9.84 mg/g, respectively, for the hybrid material *L.C* + 4% Zn^{2+} synthesized under AC, the hybrid material *L.C* + 4% Zn^{2+} synthesized in absence of AC and the pure *L.C*. The Q_m increased greatly when adding the 4% of Zn^{2+} compared to the pure *L.C*. These findings can be explained by the high dispersion of active sites due to the availability of a large number of sites. The modification of pure *L.C* by different zinc rates in the presence and absence of AC increased the active sites available for MB retention. Furthermore, the rapid change in current direction has enhanced the access of pollutants into the biosorbent pores—which explained the main difference between rates achieved in both cases.

Table 5. Comparison of biosorption characteristics between hybrid material synthesized and other studied biosorbents used for MB biosorption.

Biosorbents/MB	Q_m (mg/g)	References
Luffa cylindrica	3.9	This study
L.C + 1% Zn^{2+}	5.47	This study
L.C + 2% Zn^{2+}	6.55	This study
L.C + 4% Zn^{2+}	8.77	This study
L.C + 1% Zn^{2+} + AC	6.06	This study
L.C + 2% Zn^{2+} + AC	7.33	This study
L.C + 4% Zn^{2+} + AC	9.01	This study
Algae Caolina	6	[57]
Langsat peel	8	[58]
Agave Americana	6	[59]
Posidonia Oceanica	5.51	[60]
Phragmites Australis	6.31	[61]

3.6. Evaluation of the Synthesized Hybrid Material on the Purification of Industrial Textile Wastewater

Some experiments were performed in order to foresee the variation of pH, COD, turbidity, TOC and processing time during the purification of industrial wastewater (Table 6). In fact, the treated wastewater is composed of sodium carbonate, caustic soda, sodium sulfate, sea salt, wetting agent, reactive dyes, chlorine, and enzymes. *L.C* + 4% Zn^{2+} synthesized in the presence and absence of AC and the pure *L.C* were evaluated for the purification.

Table 6. Comparative results obtained for the purification of industrial textile wastewater.

Parameters	Industrial Textile Wastewater		
Biosorbent Used	Pure *L.C*	*L.C* + 4% Zn^{2+}	*L.C* + 4% Zn^{2+} + AC
pH_i	11.22	11.22	11.22
pH_f	8.99	8.08	8
COD_i (mg/L)O_2	1265	1265	1265
COD_f (mg/L)O_2	597	297	159
COD removal (%)	52.80	76.5	87.43
Turbidity$_i$ (NTU)	11.70	11.70	11.70
Turbidity$_f$ (NTU)	4.58	2.23	1.99
Turbidity removal (%)	60.85	80.94	82.99
TOC_i (ppm(C.O))	28.9	28.9	28.9
TOC_f (ppm(C.O))	11.9	3.2	2.29
TOC removal (%)	58.82	88.92	92.07
Processing time (min)	300	240	180

Results show that the hybrid material (*L.C* + 4% Zn^{2+}) synthesized under AC is more efficient when synthesized in the absence of AC and pure *L.C* for purification of industrial textile wastewater.

The % of TOC, % of COD and the % of turbidity removals obtained for the hybrid material synthesized in presence of AC were higher than those obtained by hybrid material *L.C* + 4% Zn^{2+} synthesized in absence of AC and pure *L.C.*

Both 87.43% of COD and 92.07% of TOC were removed from the industrial textile wastewater when using the hybrid material *L.C* + 4% Zn^{2+} synthesized in presence of AC. 52.80% of COD and 58.82% of TOC were removed from the tested industrial wastewater when using the pure *L.C.* 76.5% of COD and 80.94% of TOC were removed when using the hybrid material *L.C* + 4% Zn^{2+} synthesized in absence of AC, for the purification of the industrial wastewater.

The same behavior was observed for the % of turbidity removal. Values reached about 60.85 for pure *L.C,* 80.94% for the hybrid material (*L.C* + 4% Zn^{2+}) synthesized in absence of AC and 82.99%, respectively, for hybrid material (*L.C* + 4% Zn^{2+}) synthesized in presence of AC.

While for pH, both *L.C* and hybrid materials (*L.C* + 4% Zn^{2+}) had a significantly poor effect. This considerable fluctuation in the evaluated parameters may be attributed to the increase of the sites available for the color retention offered by the elaborated hybrid material as indicated by the Brouers-Sotolongo modeling. This analysis revealed that almost all of the parameters satisfied the Tunisian standards TN-106-02 indicating that the treated wastewater can be discharged to any receiving water body.

3.7. Phytotoxicity Test

In order to evaluate the suitability of the raw and treated industrial textile wastewater for agricultural use, a phytotoxicity test was done. The evolution of the germination index (GI), the length of root, stem and leaf of lettuce using distilled water, raw and treated industrial textile wastewater was evaluated. The most used model for weeds is lettuce *"Lactuca Sativa"*. It has been used extensively thanks to its fast germination and high sensitivity for a large variety of pollutants. Furthermore, it is often used for the examination of plants interaction in an aquatic environment [62,63].

Figure 6 (Figure S1 in the complementary information) presents the main results obtained. The germination index (GI) of the initial concentrations of the tested industrial wastewater was equal to 29.87%. After biosorption assisted by AC, the percentage of the germination index increased indicating a decrease in the toxicity of the wastewater tested. It reached respectively 72.33, 59.79, and 52.14%, respectively, when using the hybrid material (*L.C* + 4% Zn^{2+}) synthesized in presence of AC, the hybrid material *L.C* + 4% Zn^{2+} synthesized in presence of AC and the pure *L.C*. Results showed that all GI were higher than 50%, indicating the purity of the treated wastewater and its suitability to agricultural use.

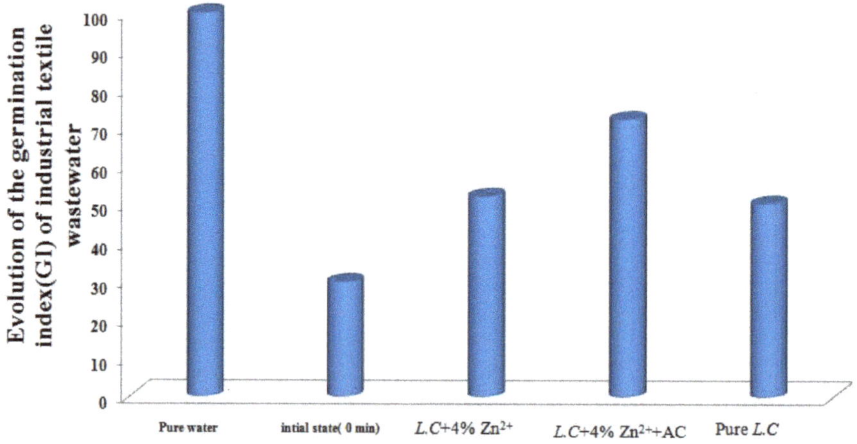

Figure 6. Evolution of the % GI after coupling biosorption with AC using pure *L.C* and hybrid material *L.C* + 4% Zn^{2+} synthesized in presence and absence of AC.

3.8. Mechanism of Biosorption of MB

An assumption of the pathway of the fixation of Zn^{2+} ions onto the *L.C* fibers and the possible interactions between MB and the synthesized hybrid material were predicted as shown in Figures 7 and 8.

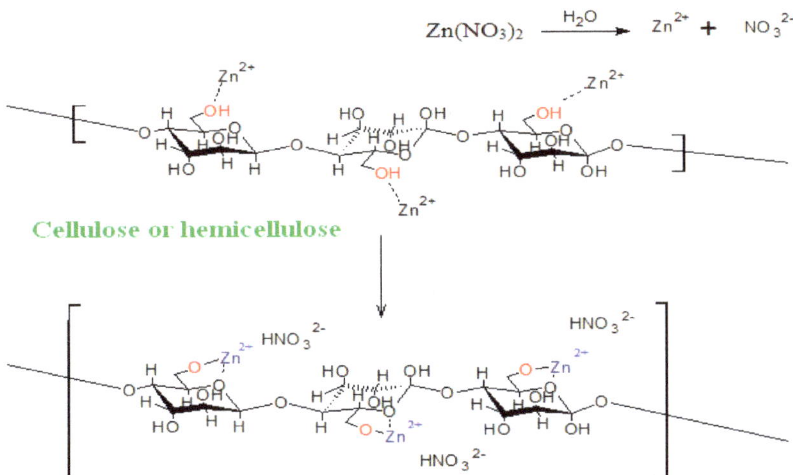

Figure 7. Possible pathway for the fixation of Zn^{2+} onto the *L.C* surface.

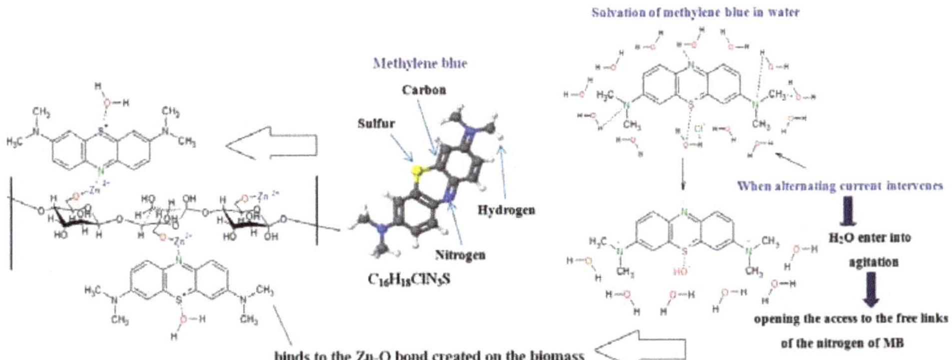

Figure 8. Possible pathway for the interactions between methylene blue and the synthesized hybrid material.

Figure 7 (Figure S2 in the complementary information) presents the possible pathway of the fixation of Zn^{2+} ions onto the *L.C* fibers in the presence and absence of AC.

For both methods used the Zn^{2+} ions generated from the dissolution of $Zn(NO_3)_2$ in water were fixed on the surface of the *L.C* through cellulose or hemicellulose, where oxygen is found, ensuring the creation of Zn-O bond on the biomass as confirmed by IR characterizations. The use of AC has enhanced the access of the Zn^{2+} ions onto the fibers. Therefore, favoring the participation of all Zn^{2+} ions presents in solution in the creation of Zn-O and Zn-O-C bonds. These findings can be confirmed by IR characterizations. The band located at 1383 cm^{-1} appears from 1% of Zn^{2+} for the hybrid material synthesized under AC, while it appears only for 4 % of Zn^{2+} for the hybrid materials synthesized in absence of AC ensuring an increase in the number of sites available onto the hybrid material surface.

Figure 8 (Figure S3 in the complementary information) presents the proposal of a possible pathway of the interactions between MB and the synthesized hybrid material and the effect of alternating current on the proposal process.

Water solvates MB by making hydrogen bridges hydrogen obtained from and nitrogen obtained from MB and a polar bond between oxygen and sulfur (π–π Bond).

When alternating current intervenes, the molecules of H_2O enter into agitation opening the access to the free links of the nitrogen of MB which subsequently binds to the Zn-O bond created on the biomass.

4. Conclusions

We have successfully synthesized a performing hybrid material based on *Luffa cylindrica* fibers and different percentage of zinc oxide (1%, 2%, and 4%) by an easy precipitation method under AC. The fast and efficient pollutant removals are the basic benefits of this research. Results confirmed clearly the effect of AC on the modification of acidic-basic properties of the *L.C* compared to the pure one. The addition of the different percentage of Zn^{2+} (1%, 2%, and 4%) has increased the basicity of the surface functionality by increasing the pH_{pzc}. Boehm titration indicated that the use of AC in the preparation of the hybrid materials has decreased the carboxylic and phenolic groups and increased the lactonic group. The presence of more lactonic groups has increased the density of the negative charge on the hybrid material surface which will play a crucial effect on the positively charged dye (MB) biosorption capacity. AC has increased the fixation and the distribution of Zn^{2+} into the *L.C* fibers by increasing the sites' number as confirmed by B.S (1, α). Physicochemical analyses revealed a prominent decrease in the COD, the TOC and the turbidity of the treated industrial textile wastewater, complied with the Tunisian Standards TN-106-02. The treated textile wastewater can be discharged to any receiving water. Furthermore, the germination indexes are higher than 50% confirmed their suitability to agricultural use.

Supplementary Materials: The following are available online at http://www.mdpi.com/2073-4441/11/7/1326/s1, Figure S1: Evolution of the % GI after coupling biosorption with AC using pure *L.C* and hybrid material *L.C* + 4% Zn^{2+} synthesized in presence and absence of AC, Figure S2: Possible pathway for the fixation of Zn^{2+} onto the *L.C* surface, Figure S3: Possible pathway for the interactions between methylene blue and the synthesized hybrid material, Table S1: Comparative results for the B.S model.

Author Contributions: Conceptualization, A.O., A.K. and M.S.; Formal analysis, A.O., A.K. and M.S.; Methodology, A.O.; resources, A.O., A.K., M.S. and R.B.; Writing—Original Draft preparation, A.O.; Writing—Review and Editing, A.O., A.K., M.S., R.B. and M.V.

Funding: This study was supported by 5TOI_EWAS project funded by the European Union's Horizon 2020 Research and Innovation Program under Grant Agreement No. 692523.

Acknowledgments: The authors express their sincere thanks to the Laboratory of Energy and Materials (High School of Sciences and technology of Hammam Sousse)for the financial support and the Unitat de QuímicaAnalítica, Departament de Química, Centre Grup de Técniques de Separacio'en Química (GTS), Universitat Autónoma de Barcelona, Bellaterra, Spain.

Conflicts of Interest: The authors declare no conflict of interest.

References

1. Adegoke, K.A.; Bello, O.S. Dye sequestration using agricultural wastes as adsorbents. *Water Resour. Ind.* **2009**, *12*, 8–24. [CrossRef]
2. Laasri, L.; Elamrani, M.K.; Cherkaoui, O. Removal of two cationic dyes from a textile effluent by filtration-adsorption on woodsawdust. *Environ. Sci. Pollut. Res.-Int.* **2007**, *14*, 237–240. [CrossRef] [PubMed]
3. Abidin, F.C.Z.A.; Rahmat, N.R. Multi-stage ozonation and biological treatment for removal of azo dye industrial effluent. *Int. J. Environ. Sci. Dev.* **2010**, *1*, 193–198.
4. Othmani, A.; Kesraoui, A.; Seffen, M. The alternating and direct current effect on the elimination of cationic and anionic dye from aqueous solutions by electrocoagulation and coagulation flocculation. *Euro-Mediterr. J. Environ. Integr.* **2017**, *2*, 2–6. [CrossRef]
5. Selmi, T.; Sanchez-Sanchez, A.; Gadonneix, P.; Jagiello, J.; Seffen, M.; Sammouda, H.; Celzard, A.; Fierro, V. Tetracycline removal with activated carbons produced by hydrothermal carbonisation of Agave americana fibres and mimosa tannin. *Ind. Crop. Prod.* **2018**, *115*, 146–157. [CrossRef]

6. Căilean, D.; Barjoveanu, G.; Musteret, C.P.; Sulitanu, N.; Manea, L.R.; Teodosiu, C. Reactive dyes removal from wastewater by combined advanced treatment. *Environ. Eng. Manag. J.* **2009**, *8*, 503–511. [CrossRef]

7. Vasudevan, S.; Lakshmi, J.; Sozhan, G. Effects of alternating and direct current in electrocoagulation process on the removal of cadmium from water. *J. Hazard. Mater.* **2011**, *192*, 26–34. [CrossRef] [PubMed]

8. Vijayageetha, V.A.; Rajan, A.P.; Arockiaraj, S.P.; Annamalai, V.; Janakarajan, V.N.; Balaji, M.D.S.; Dheenadhayalan, M.S. Treatment study of dyeing industry effluents using reverse osmosis technology. *Res. J. Recent Sci.* **2014**, *3*, 58–61.

9. Elaissaoui, I.; Akrout, H.; Grassini, S.; Fulginiti, D.; Bousselmi, L. Role of SiOx interlayer in the electrochemical degradation of Amaranth dye using SS/PbO$_2$ anodes. *Mater. Des.* **2016**, *10*, 633–643. [CrossRef]

10. Vendruscolo, F.; Ferreira, G.L.R.; Filho, N.R.A. Biosorption of hexavalent chromium by microorganisms. *Int. Biodeterior. Biodegrad* **2017**, *119*, 87–95. [CrossRef]

11. Davis, T.A.; Volesky, B.; Mucci, A. A review of the biochemistry of heavy metal biosorption by brown algae. *Water Res.* **2003**, *37*, 4311–4330. [CrossRef]

12. Atar, N.; Olgun, A.; Colk, F. Thermodynamic, Equilibrium and Kinetic Study of the Biosorption of Basic Blue 41 using *Bacillus maceran*. *Eng. Life Sci.* **2008**, *8*, 499–506. [CrossRef]

13. Champenois, W.; Borges, A.V. Determination of dimethyl sulfonio propionate and dimethyls ulfoxide in Posidonia oceanic leaf tissue. *MethodsX* **2019**, *6*, 56–62.

14. Mbarki, F.; Kesraoui, A.; Seffen, M.; Ayrault, P. Kinetic, thermodynamic and adsorption behavior of cationic and anionic dyes onto Corn stigmata: Non-linear and stochastic analyses. *Water Air Soil* **2018**, *229*, 95. [CrossRef]

15. Kesraoui, A.; Mabrouk, A.; Seffen, M. Valuation of Biomaterial: *Phragmites australis* in the Retention of Metal Complexed Dyes. *Am. J. Environ. Sci.* **2017**, *3*, 266–276. [CrossRef]

16. Demir, H.; Top, A.; Balkose, D.; Ulk, S. Dye adsorption behavior of Luffa cylindrica fibers. *J. Hazard. Mater.* **2008**, *153*, 389–394. [CrossRef] [PubMed]

17. Wang, S.; Boyjoo, Y.; Choueib, Z.H.A. Removal of dyes from aqueous solution using fly ash and red mud. *Water Res.* **2015**, *39*, 129–138. [CrossRef]

18. Hachani, R.; Sabir, N.S.; Zohra, K.F.; Nesrine, M.N. Performance Study of a Low-cost Adsorbent-Raw Date Pits-for Removal of Azo Dye in Aqueous Solution. *Water Environ. Res.* **2017**, *98*, 827–839. [CrossRef]

19. El Shahawy, A.; Ghada Heikal, G. Organic pollutants removal from oily wastewater using clean technology economically, friendly biosorbent (Phragmites australis). *Ecol. Eng.* **2018**, *122*, 207–218. [CrossRef]

20. Osman, C. Synthèse De Nouveaux Matériaux Hybrides Pour Les Catalyses En Atrp Supporté Du Méthacrylate De Méthyle. Ph.D. Thesis, University of Pierre and Marie Curie, Paris, France, 2014.

21. Shivaramakrishnan, B.; Gurumurthy, B.; Balasubramanian, A. Potential biomedical applications of metallic nanibiomaterials: A review. *Int. J. Pharm. Sci. Res.* **2017**, *8*, 98500.

22. Ye, J.D.; Gu, S.L.; Qin, F.; Zhu, S.M.; Liu, S.M.; Zhou, X.; Liu, W.; Hu, L.Q.; Zhang, R.; Shi, Y.; et al. MOCVD growth and properties of ZnO films using dimethylzinc and oxygen. *Appl. Phys. A* **2005**, *81*, 809–812. [CrossRef]

23. Tului, M.A.; Bellucci, A.; Migliozzi, G. Zinc oxide targets for magnetron sputtering PVD prepared by plasma spray. *Surf. Coat. Technol.* **2005**, *205*, 1070–1073. [CrossRef]

24. Zamiri, R.; Zakaria, H.A.; Ahangar, M.; Zak, A.K.; Drummen, G.P.C. Aqueous starch as a stabilizer in zinc oxide nanoparticle synthesis via laser ablation. *J. Alloys Compd.* **2012**, *516*, 41–48. [CrossRef]

25. Perelshtein, L.; Applerot, G.; Perkas, N.; Wehrschetz-Sigl, E.; Hasmann, A.; Guebitz, G.M.; Gedanken, A. Antibacterial Properties of an In Situ Generated and Simultaneously Deposited Nanocrystalline ZnO on Fabrics. *ACS Appl. Mater. Interfaces* **2008**, *1*, 361–366. [CrossRef] [PubMed]

26. Hu, W.; Chen, S.; Zhou, B.; Wang, H. Facile synthesis of ZnO nanoparticles based on bacterial cellulose. *Mater. Sci. Eng. B* **2010**, *170*, 88–92. [CrossRef]

27. Ching, C.G.; Lee, S.C.; Ooi, P.K.; Ng, S.S.; Hassan, Z.; Hassan, H.A.; Abdullah, M.J. Optical and structural properties of porous zinc oxide fabricated via electrochemical etching method. *Mater. Sci. Eng. B* **2013**, *178*, 956–959. [CrossRef]

28. Chen, C.Y.; Chang, H.W.; Shih, S.J.; Tsay, C.Y.; Chang, C.J.; Lin, C.K. High super capacitive stability of spray pyrolyzed ZnO-added manganese oxide coatings. *Ceram. Int.* **2013**, *39*, 1885–1892. [CrossRef]

29. Brinker, C.J.; Scherer, G.W. *Sol-Gel Science: The Physics and Shemistry of Sol-Gel Processing Academic*; ET Press: Frankfurt, Germany, 1990.

30. Kesraoui, A.; Bouzaabia, S.; Seffen, M. The combination of Luffa cylindrical fibers and metal oxides offers a highly performing hybrid fiber material in water decontamination. *Environ. Sci. Pollut. Res.* **2018**, *26*, 11524–11534. [CrossRef]

31. Kesraoui, A.; Moussa, A.; Ali, G.B.; Seffen, M. Biosorption of alpacide blue from aqueous solution, by lignocellulosicbiomass: *Luffa cylindrical* fibers. *Environ. Sci. Pollut. Res.* **2015**, *23*, 15832–15840. [CrossRef]

32. Shafei, A.; Nikzar, M.; Arami, M. Photocatalytic degradation of terephlatic acid using titania and zinc oxide ZnO photocalyts comparative study. *Desalination* **2013**, *252*, 8–16. [CrossRef]

33. Kesraoui, A.; Selmi, T.; Seffen, M.; Brouer, F. Influence of alternating current on the adsorption of indigo carmine. *Environ. Sci. Pollut. Res.* **2016**, *24*, 9940–9950. [CrossRef] [PubMed]

34. Boumediene, M.; Benaïssa, H.; George, B.; Molina, S.; Merlin, A. Effects of pH and ionic strength on methylene blue removal from synthetic aqueous solutions by sorption onto orange peel and desorption study. *J. Mater. Environ. Sci.* **2018**, *9*, 1700–1711.

35. Goertzen, L.; ThériaultKim, S.D.; Oickle, M.A.; Tarasuk, C.A.; Andreas, A.H. Standardization of the Boehm titration. Part I. CO_2 expulsion and endpoint determination. *Carbon* **2010**, *48*, 1252–1261. [CrossRef]

36. Nava, J.L.; Quiroz, M.A.; Martinez-Huitle, C.A. Electrochemical treatment of synthetic wastewaters containing Alphazurine A dye: Role of electrode material in the color and COD removal. *J. Mex. Chem. Soc.* **2008**, *52*, 249–255.

37. Sharma, A.; Bhattacharya, S.; Sen, R.; Reddy, B.S.B.; Fecht, H.J.; Das, K.; Das, S. Influence of current density on microstructure of pulse electrodeposited tin coatings. *Mater. Charact.* **2012**, *68*, 22–32. [CrossRef]

38. Amrhar, O.; Nassali, H.; Elyoubi, M.S. Adsorption of a cationic dye, Methylene Blue, ontoMoroccan Illitic Clay. *J. Mater. Environ. Sci.* **2015**, *6*, 3054–3065.

39. Brouers, F. Statistical foundation of empirical isotherms. *Open J. Stat.* **2014**, *4*, 687–701. [CrossRef]

40. Brouers, F. The fractal (BSf) kinetics equation and its approximations. *J. Mod. Phys.* **2014**, *5*, 1594–1601. [CrossRef]

41. Brouers, F.; Al-Musawi, T.J. On the optimal use of of isotherm models for the characterization of biosorption of lead onto algae. *J. Mol. Liq.* **2015**, *212*, 46–51. [CrossRef]

42. Khot, L.R.; Sankaran, S.; Maja, J.M.; Ehsani, R.; Schuster, E.W. Applications of nanomaterials in agricultural production and crop protection: A review. *Crop Prot.* **2012**, *35*, 64–70. [CrossRef]

43. Antonio, R.; Tapia-Benavides, A.R.T.; Tlahuextl, M.; Tlahuext, H.; Galán-Vidala, C. Synthesis of Zn compounds derived from *1H*-benzimidazol-2-ylmethanamine. *Arkivoc* **2008**, *5*, 172–186.

44. Xu, Q.; Chena, C.; Rosswurma, K.; Yaoa, T.; Janaswamy, S. A facile route to prepare cellulose-based films. *Carbohydr. Carbohydr. Polym.* **2016**, *149*, 274–281. [CrossRef]

45. Vergnat, V. Matériaux Hybrides Organique-Inorganique Par Greffage Covalent De Polymères Sur Des Oxydes Métalliques. Ph.D. Thesis, University of Strasbourg, Strasbourg, France, 2011.

46. Zeleňák, V.; Vargová, Z.; Györyová, K. Correlation of infrared spectra of zinc (II) carboxylates with their structures. *Spectrochim. Acta A Mol. Biomol. Spectrosc.* **2007**, *66*, 262–272. [CrossRef]

47. Li, L.; Deng, J.; Deng, H.; Liu, Z.; Li, X. Preparation, characterization and antimicrobial activities of chitosan/Ag/ZnO blend films. *Chem. Eng. J.* **2010**, *160*, 378–382. [CrossRef]

48. Liu, X.F.; Sun, J.; Ren, J.; Curling, S.; Sun, R. Physicochemical characterization of cellulose from perennial leaves (Loliumperenne). *Carbohydr. Res.* **2006**, *341*, 2677–2687. [CrossRef] [PubMed]

49. Awala, H.; El-Roz, M.; Toufaily, J.; Mintova, S. Template-Free Nanosized EMT Zeolite Grown on Natural Luffa Fibers for Water Purification. *Adv. Porous Mater.* **2014**, *4*, 141–182. [CrossRef]

50. Arefi, M.R.; Zarchi, S.R. Synthesis of Zinc Oxide Nanoparticles and Their Effect on the Compressive Strength and Setting Time of Self-Compacted Concrete Paste as Cementitious Composites. *Int. J. Mol. Sci.* **2012**, *13*, 4340–4350. [CrossRef]

51. Wan, C.; Jian, L. Embedding ZnO nanorods into porous cellulose aerogels via a facile one-step low-temperature hydrothermal method. *Mater. Des.* **2015**, *83*, 620–625. [CrossRef]

52. Mahmoudi, K.H.; Hamdi, N.; Srasra, E. Kinetics and equilibrium studies on removal of methylene blue and methyl orange by adsorption onto activated carbon prepared from date pits—A comparative study. *Water Air Soil Pollut.* **2018**, *229*, 95. [CrossRef]

53. Cenens, J.; Schoonheydt, R.A. Visible spectroscopy of methylene blue on hectorite, laponite and barasym in aqueous suspension. *Clays Clay Min.* **1988**, *36*, 214–224. [CrossRef]

54. Kalmár, J.; Lente, G.; Fabian, I. Kinetics and mechanism of the adsorption of methylene blue from aqueous solution on the surface of a quartz cuvette by on-line UV-Vis spectrophotometry. *J. Dyes Pigments.* **2016**, *127*, 170–178. [CrossRef]

55. Selmi, T.; Seffen, M.; Sammouda, H.; Mathieu, S.; Jagiello, J.; Celzard, A.; Fierro, V. Physical meaning of the parameters used in fractal kinetic and generalised adsorption models of Brouers–Sotolongo. *Adsorption* **2017**, *24*, 11–27. [CrossRef]

56. Hammud, H.H.; Fayoumi, L.M.A.; Holail, H.; Mostafa, E.S.M.E. Biosorption studies of methylene blue by Mediterranean algae Carolina and its chemically modified forms. Linear and nonlinear models' prediction based on statistical error calculation. *Int. J. Chem.* **2011**, *3*, 147–163. [CrossRef]

57. Salleh, M.A.M.; Mahmoud, D.K.; Binti Awang Abu, N.A.; Abdul Karim, W.A.W.; Idris, A.B. Methylene blue adsorption from aqueous solution by langsat (lansium domesticum) peel. *J. Purity Util. React. Environ.* **2012**, *1*, 442–465.

58. Hamissa, A.M.B.; Brouers, F.; Ncibi, M.C.; Seffen, M. Kinetic Modeling Study on Methylene Blue Sorption onto Agave americana fibers: Fractal Kinetics and Regeneration Studies. *Sep. Sci. Technol.* **2013**, *48*, 2834–2842. [CrossRef]

59. Ncibi, M.C.; Mahjoub, B.; Seffen, M. Studies on the biosorption of textile dyes from aqueous solutions using Posidonia Oceanic (L.) leaf sheaths fibres. *Adsorpt. Sci. Technol.* **2006**, *24*, 461–473. [CrossRef]

60. García-Ávila, F.; Patiño-Chávez, J.; Zhinín-Chimbo, F.; Donoso-Moscoso, S.; del Pino, L.F.; Avilés-Añazco, L. Performance of Phragmites Australis and Cyperus Papyrus in the treatment of municipal wastewater by vertical flow subsurface constructed wetlands. *Int. Soil Water Conserv. Res.* 2019. [CrossRef]

61. Dallel, D.; Kesraoui, A.; Seffen, M. Biosorption of cationic dye onto "Phragmites australis" fibers: Characterization and mechanism. *J. Environ. Chem. Eng.* **2018**, *6*, 6. [CrossRef]

62. Louhichi, G.; Bousselmı, L.; Ghrabi, A.; Khouni, I. Process optimization via response surface methodologyin the physico-chemical treatment of vegetable oil refinery wastewater. *Environ. Sci. Pollut. Res.* **2018**, 1–19. [CrossRef]

63. Mendes, P.M.; Becker, R.; Corrêa, L.B.; Bianchi, I.; Dai Prá, M.A.; Lucia, T., Jr.; Corrêa, E.K. Phytotoxicity as an indicator of stability of broiler production residues. *J. Environ. Manag.* **2016**, *167*, 156–159. [CrossRef]

Article

Filtration of Uncharged Solutes: An Assessment of Steric Effect by Transport and Adsorption Modelling

Simona M. Miron [1,2], Patrick Dutournié [1,2,*] and Arnaud Ponche [1,2]

[1]   Institut de Science des Matériaux de Mulhouse IS2M (UMR CNRS 7228), Université de Haute Alsace,
      68100 Mulhouse, France; simona-melania.miron@uha.fr (S.M.M.); arnaud.ponche@uha.fr (A.P.)
[2]   Université de Strasbourg, 3 bis rue A. Werner, 68093 Mulhouse CEDEX, France
*   Correspondence: patrick.dutournie@uha.fr; Tel.: +33-389-336-752

Received: 3 September 2019; Accepted: 12 October 2019; Published: 19 October 2019

Abstract: The major aim of this work was to understand and estimate the evolution of the membrane selectivity of neutral solutes after the filtration of protein or amino acid solutions. Classical methodologies led to the estimation of the mean pore radius, different for each filtrated neutral solute. The use of pore size distribution from nitrogen adsorption/desorption experiments enabled a good description of hydraulic and selectivity performances. The modification of the membrane hydraulic properties after the successive filtration of protein solutions revealed that the decrease is quasi linear, the same for all the studied membranes and independent of prior tests. According to the experimental observations, an adsorption model was developed, considering a layer by layer adsorption in the larger pores of the membrane. The predictive obtained results are in good agreement with the experimental rejection rates, validating the assumptions.

Keywords: protein adsorption; neutral solute; ultrafiltration; selectivity modelling; pore size distribution

1. Introduction

Membrane-based technology offers a reliable option in a growing market to the standard separation processes (especially in terms of power consumption, addition of chemical reagents, and operating convenience) [1]. Currently, membrane separation technology is widely used to separate, concentrate, and rectify solutions for various purposes (for example, waste treatment, water desalination, purification of pharmaceuticals) in many industrial sectors (food, medicine, waste decontamination, chemical, textile industries) [2–5]. Specifically, they are used for the separation of neutral solutes from aqueous solutions (proteins for food industry, vitamins, peptides, drugs in general for the medicine industry) [6,7].

Alternatively, neutral solute filtrations are mainly used for characterizing membranes (steric effect, cut-off). For this purpose, a transport model is used. For example, Schönherr et al. [8] filtrated raffinose to estimate the mean pore radius by using the Paganelli and Solomon model [9]. Schaerp et al. [10] used the Spiegler–Kedem equation [11] derived from irreversible thermodynamics to numerically approximate the experimental rejection curves of galactose–, maltose–, and raffinose–water solutions. The reflection coefficient is expressed by the steric hindrance pore model. Ito et al. [12] filtrated five dextran samples (molecular weights ranging from 10.5 kg mol^{-1} to 2000 kg mol^{-1}) to investigate the modification of pore size as a function of the Ba^{2+} concentration. Velicangil and Howell [13] estimated the steric properties of ultrafiltration membranes by using the orifice model with rejection curves of three protein solutions (papain, Bovine Serum Albumin (BSA), and ovalbumin). In recent years, several works investigated the steric effect by using the Nernst–Planck model for uncharged solutes [14]. They studied the rejection rates of vitamin B12, raffinose, sucrose, and glucose and compared the estimated mean pore radius with the results from Atomic Force Microscopy (AFM). Significant differences were observed between the estimated mean pore radius, especially regarding pore size distribution estimated via AFM investigations. This methodology was used in many current works [15–17].

One major observation of these works is that results obtained with several molecules are different. Indeed, the filtration of different uncharged solutes leads to various estimations of mean pore radius. Another major problem, mainly related in the literature, is the adsorption of uncharged solutes in the membrane pore, and so, the partial clogging of the active layer. Proteins are indeed known to adsorb on various surfaces due to their high propensity of deformability and conformation modification, as demonstrated by Karlson et al. [18] and Norde [19]. Even if the amount of protein on hydrophobic substrates is generally high, the outer layer of membranes constituting of oxides can also absorb a non-negligible quantity of protein. Thereby, it reduces the apparent pore size and also modifies the surface properties of the active layer, as demonstrated by Huang et al. [20] and Ponche et al. [21].

For example, Robertson and Zydney [22] observed protein adsorption in ultrafiltration membrane pores, reducing the hydraulic permeability and increasing the selectivity. They estimated the decrease in pore size due to adsorption, a decrease compatible with a monolayer adsorption. A.M. Comerton et al. [23] studied the membrane adsorption of pharmaceutically active compounds. They observed that ultrafiltration (UF) membranes are more subject to adsorption than nanofiltration and reverse osmosis membranes. This phenomenon can be easily explained by the pore size of UF membranes. In most cases, the adsorption of uncharged solutes increases the membrane selectivity and decreases the hydraulic performances and, in some cases, can extend to complete pore clogging [24]. If studies referring to molecule adsorption in membrane pore are abundant [25], fewer studies deal with adsorption mechanisms in the pore, except the classical theoretical models (monolayer) [26] or for microfiltration membranes [27]. In any case, there has not been a study that dealt with the numerical estimation of solute rejection rate (predictive) by taking into account pore size distribution and pore adsorption.

In this study, the filtration of neutral solutes is performed with a ceramic ultrafiltration membrane to investigate size selectivity. To this end, the experimental pore size distribution estimated from nitrogen adsorption/desorption experiments is used as an input of the model. Adsorption phenomena are taken into account to understand the change in hydraulic performances and the membrane selectivity.

2. Materials and Methods

Experiments were performed with tubular ceramic membranes provided by TAMI Industries (Nyons, France). The active layer in TiO_2 was deposited on the internal surface of an alumina macroporous tube (1 kDa, length = 25 cm, inner diameter = 8 mm). The TiO_2 layer was observed by scanning electron microscopy (Philips XL30 First FEG, SEMTech Solutions, North Billerica, MA, USA) to estimate layer thickness.

Nitrogen gas adsorption–desorption isotherms were performed with a Micromeritics ASAP 2420 (Micrometrics, Ottawa, Canada, apparatus at T = 77 K. Prior to experiments, each sample (broken pieces of TiO_2 membrane and alumina tube) was out gassed to a residual pressure lower than 0.8 Pa at 350 °C for 15 h. The micropore volume was calculated using the t-plot method. The cumulative volume and the pore size diameter (distribution) were calculated using the density functional theory method (DFT).

Filtration tests were performed in a laboratory pilot plant (stainless steel), previously described in literature [28]. The studied solution was stored in a 5 L tank and a volumetric pump provided solution circulation for tangential flow filtration (retentate). The flow was controlled by an analog flow rate sensor. The experiments were performed at 700 L/h, a value corresponding to a mean fluid velocity higher than 5 m/s to avoid concentration polarization at the surface of the active layer. After filtration, the permeate flow was sampled for analysis. Both retentate and permeate returned into the solution tank. The applied pressure was adjusted by a manual valve (4 to 12 bar) and was measured by two sensors upstream and downstream the membrane carter. A cooling unit maintained the fluid temperature at 25 °C. Between each filtration test, the experimental set-up was rinsed with demineralized water (conductivity < 0.1 µS/cm).

Before performing filtration tests, a conditioning step of the membrane was required. This step consisted of the filtration of pure water until hydraulic performances were in steady state.

Filtration tests of uncharged solute-water solution ware performed with vitamin B12 (Alfa Aesar, purity 98%), L-phenylalanine (Fluka, purity 99%), L-tyrosine (Fluka, purity 99%), and lysozyme (Sigma-Aldrich, from chicken egg white, activity > 70,000 U/mg). The concentrations of retentate (C_r) and permeate (C_p) samples were investigated by absorbance measurements with a UV-visible spectrophotometer (Lambda 35, Perkin Elmer Instrument). Information about the studied molecules are given in Table 1.

Table 1. Information on the molecules used for experimental tests.

Products	Concentration (mol m^{-3})	Molar Mass (g mol^{-1})	Stokes Radius (Å)	Wavelength (nm)
L-phenylalanine	5.45	165.19	3.7	257
L-tyrosine	2.48	181.19	3.8	275
Vitamin B12	9.22×10^{-3}	1355.38	7.4	362
Lysozyme	0.025	14,300	19.0	281

The observed rejection rate R was calculated using Equation (1):

$$R = \frac{C_r - C_p}{C_r}. \tag{1}$$

Several tests of analytical measurement uncertainties were performed (repeated eight times). The maximal error was 1.9% for retentate and 2.9% for permeate samples. The rejection rate uncertainty was $\Delta R = (1-R)\left[\frac{1}{C_p}\Delta C_p + \frac{1}{C_r}\Delta C_r\right]$ and, in all cases, was inferior to 4.8% (result for a rejection rate close to 0).

During the experimental tests, the hydraulic properties were monitored. To do this, between each experiment test, a pure water filtration test was performed. The permeate flux (Jv) was measured and plotted for different applied pressures (ΔP). The hydraulic permeability (L_p) was obtained from the slope of the linear curve via Equation (2):

$$Jv = \frac{L_p}{\mu}\Delta P. \tag{2}$$

Three series of tests were performed with the same membrane. Between each series the membrane was regenerated. The regeneration consisted of a hydrothermal treatment (five days in water at 105 °C) to recover its original hydraulic properties.

Filtration tests of neutral solutes were used to investigate size selectivity. Two models were studied: model A assumed a uniform one size distribution (average pore radius) and the second (model B) used an experimental size distribution determined from nitrogen adsorption/desorption experiment as an input.

In model A, the mean pore radius was estimated by numerically approximating rejection rates of the studied solutes. The equation used to approximate the experimental results was the solution of the mass balance (differential equation) in the membrane pore. The differential equation expressing solute mass balance is described by the Nernst–Planck approach for uncharged solutes [29,30], assuming uniform dispersed, one size cylindrical pores. This differential equation was solved with the equality of chemical potentials on both sides of the active layer [31] at the pore/solution interface. Starting from these assumptions, the rejection rate of an uncharged solute [32] can be calculated with Equation (3):

$$R = 1 - \frac{\varphi K_c}{1 - (1 - \varphi K_c)\exp\left(-\frac{K_c r_p{}^2 \Delta P}{8\mu K_d D_\infty}\right)}. \tag{3}$$

With $\varphi = \left(1 - \frac{r}{r_p}\right)^2$ the steric partitioning coefficient.

The second model (model B) implied that we had an experimental distribution of pore size. In this case, the permeation flow rate was assumed to be the sum of the flow rate of each pore (Equation (4)). The flow rate in the pore was calculated using Hagen Poiseuille's law:

$$Q = J_v S_m = \frac{Lp}{\mu}\Delta P \times S_m = \sum_i np(i) \times q(i) = \sum_i np(i) \times \frac{\pi r_p(i)^4}{8\mu\Delta x}\Delta P. \tag{4}$$

The rejection rate R was calculated using the Nernst–Planck approach for uncharged solutes (Equation (3)). The calculation was carried out for each pore size, summing all the contribution weighting by the pore number and the relative flow rate (Equation (5)) as follows:

$$R = \frac{1}{Q}\sum_i np(i).q(i).R(i). \tag{5}$$

3. Results

3.1. Experimental Results and Estimation of Steric Effect

Experimental tests were performed to understand the physical phenomena that act on the mass transfer mechanisms in a porous media. For this purpose, the monitoring of hydraulic and selectivity properties was required. Three series of tests were performed with the same membrane. Between each series of tests, the membrane was regenerated according to the protocol previously described in the Materials and Methods section. This regeneration aimed to recover the initial membrane properties. Table 2 provides experimental results obtained for three series in terms of maximal rejection rate (selectivity property) and hydraulic permeability (hydraulic performance). For each series, the results are in chronological order. These series were chosen as the first molecule filtrated was different (*L*-phenylalanine for series 1, vitamin B12 and Lysozyme for series 2, and *L*-Tyrosine for the last).

Table 2. Filtration experiments (chronologic order) for the three series of tests.

Molecule	R_{Max} (%)	r_p (nm)	$Lp \times 10^{14}$ $(m^3\ m^{-2})$	R_{Max} (%)	$Lp \times 10^{14}$ $(m^3\ m^{-2})$	R_{Max} (%)	$Lp \times 10^{14}$ $(m^3\ m^{-2})$
After conditioning step			6.7		6.7		6.6
L-phenylalanine	5	3.3 ± 0.30	5.5				
L-tyrosine	5	3.5 ± 0.30	4.7			5	4.8
VB12	60	1.7 ± 0.15	4.4			40	4.9
Lysozyme	93	2.5 ± 0.05	3.6			96	4.3
Lysozyme	95	2.4 ± 0.05	3.3			99	3.7
VB12	75	1.4 ± 0.10	3.4	20	6.7		
Lysozyme	98	2.2 ± 0.04	3.1	65	4.0	100	2.9
Lysozyme	99	2.0 ± 0.03	2.8	98	2.5		
VB12	81	1.2 ± 0.05	2.8			85	2.8
Lysozyme	100	<2.0	2.2	100	2.0		
VB12	87	1.0 ± 0.04	2.3	90	2.0		
VB12	86	1.0 ± 0.04	2.2				
L-phenylalanine	16	1.6 ± 0.10	2.3				
L-tyrosine						17	2.8

The first experiments (left-hand column) were performed with a new membrane. After conditioning, the hydraulic permeability was 6.7×10^{-14} m^3 m^{-2}. The four studied molecules were filtrated one after another, from the smallest to the largest. The rejection rate increased as filtration tests were performed and, at the same time, the hydraulic permeability significantly decreased, suggesting that a part of the protein was adsorbed at the membrane surface (or in the pore), reducing mass transfer. Nevertheless, filtration of VB12 does not seem to modify the membrane hydraulic performances, and, thus, was chosen as the model solute to follow the membrane selectivity properties. The solute rejection rates obtained with the membrane still having its initial properties (i.e., membrane that had only filtrated water and vitamin B12 after regeneration) were 5% for *L*-phenylalanine (test 1 series 1), 5% for *L*-Tyrosine (test 1—series 3), 20% for Vitamin B12 (test 1 series 2), and 65% for Lysozyme (test 2—series 2). These rejection rates are in good agreement with the size of the studied molecules, but much lower than expected for a membrane with a cut-off of 1 kDa.

After these first experimental tests, the membrane permeability and the solute transmission decreased. This behavior was illustrated by the rejection rate of vitamin B12, which increased as protein or amino acid solutions were filtrated. For example, it increased from 20% to 90% after several filtration tests of lysozyme solutions (series 2), indicating that steric effect increased significantly. This can be explained by protein or amino acid adsorption at the membrane surface and or in the pore, thereby restricting the transfer of solute through the porous medium. These adsorption phenomena are not completely irreversible, since after the regeneration treatment the membrane recovered its initial properties.

The estimated mean pore radii are given in Table 2 for series 1 (from Equation (3)). The results (calculated average pore radii) corroborate previous observations. Indeed, the average pore radius decreased as protein filtration was performed. The results obtained with the different solutes show that this current model (model A) remains unsatisfactory for estimating steric effect because the results are dependent on the studied molecule. The average pore radius estimated with vitamin B12 was systematically lower than for lysozyme and *L*-phenylalanine. Bowen et al. [14] obtained the same results for the filtration of four uncharged solutes (vitamin B12, raffinose, sucrose, and glucose). Indeed, the estimated mean pore radius can vary by a ratio of 2:1. They also observed a modification of water permeability after filtration of each solute.

Additional tests were performed with mixtures of uncharged solutes (vitamin B12 and lysozyme) in water or salted water. These tests were performed following the tests of series 1.

The results (Table 3) show that the rejection rate of the solute was not modified by any another solute in the solution or by salt (NaCl—5 mM) in dilute solution. In the present case, it seems that the solutes in solution did not interact between them. However, in the literature, several studies [33–35] showed an increase in neutral solute transmission in saline solutions. This phenomenon is explained by a partial dehydratation of the molecule in the membrane pore, facilitating the transport through the membrane active layer. This behavior is specifically observed in nanofiltration. In the present study, the filtrated solutes, the porous material (membrane active layer), and the pore size were different, which could explain the observed differences.

Table 3. Filtration tests of solute mixture.

Filtrated Solutions	R_{Max} (%)	Average r_p (nm)	$Lp \times 10^{14}$ (m^3 m^{-2})
Lysozyme	99	2.0	2.3
VB12 + **Lysozyme**	86 + 99	1.0/2.0	2,2
VB12 + **Lysozyme** + NaCl	87 + 100	1.0/<2.0	2,4
VB12	86	1.0	2,3

Moreover, the difference in estimated mean pore radius according to the studied molecule was confirmed. There are two possible explanations. First, the one size cylindrical pore hypothesis is a poor and inadequate representation of the porous medium. Second, the solute interacts with the membrane surface according to the chemical groups of both the solute and the active layer (van der Waals forces or acid-base interactions). To investigate the first possibility, a pore size distribution was studied. To this end, nitrogen adsorption/desorption experiments were used to measure the porous volume according to the pore size.

Figure 1 shows the cumulative porous volume obtained during the nitrogen adsorption step according to pore size. This curve can be suitably approximated by a log-normal distribution. These results were substantiated by Scanning Electron Microscopy (SEM) investigations. Indeed, the SEM images (Figure 2c) show a top view of the active layer. The surface constituted sintered titania aggregates separated by nanometric cracks or channels. The larger ones, measuring about 4–10 nm, were of a lower amount than the smaller ones (less than 1 nm wide).

From these experimental results, we assumed that the porous medium was constituted of cylindrical and unidirectional pores according to the previous size distribution (Figure 1) and calculated the pore number as a function of the pore size.

Assuming that the flow rate in each pore can be described by the Hagen–Poiseuille law, the permeation flux can be estimated with Equation (4). From these results, it is possible to estimate the active layer thickness by equaling the observed permeation flux with the flow rate calculated by Equation (4). For this purpose, the studied range of pore diameter (0–14 nm) was divided into sub-ranges of pore size from $i = 0$ to 1400. The active layer thickness was around 2.9 μm. Previous observations [28] performed by microscopy (SEM) showed that the active layer was about 1–2 μm thick. Figure 2a shows a cross section of the membrane obtained after breaking it. The different alumina layers (different porosities) are visible below the titania active layer (full grey layer). The apparent thickness (Figure 2c) of this layer is in agreement with previous investigation.

Calculations were performed for modelling selectivity performances of the new or regenerated membranes for the four studied molecules (i.e., for membranes, which have only been in contact with water). These calculated results were compared with the experimental observed rejection rates of the four studied solutes (Table 4). The results are very close to the experimental ones for the four studied molecules. This comparison indicates that taking into account pore size distribution provides a good way to estimate rejection rates of neutral solutes.

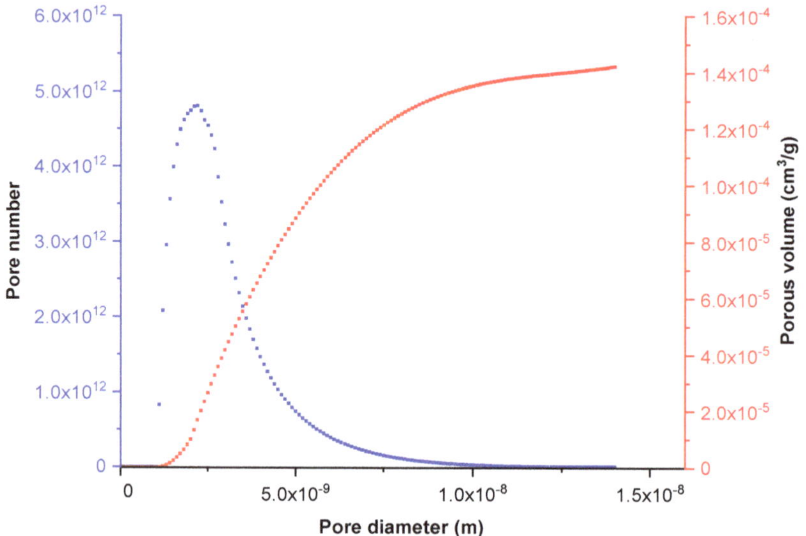

Figure 1. Cumulative porous volume (red curve) versus pore diameter and pore size distribution (blue curve).

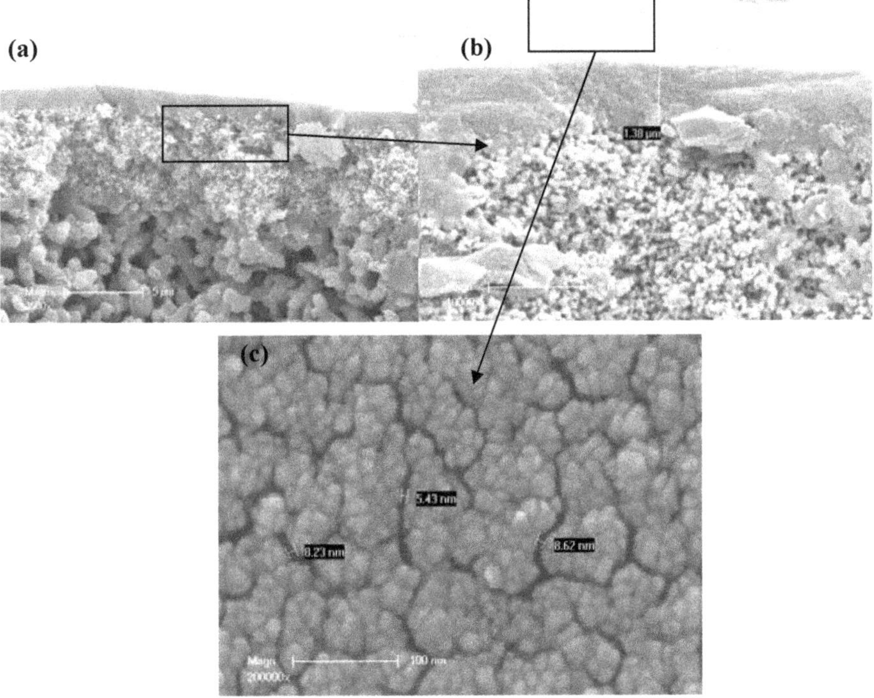

Figure 2. SEM photographs of (**a,b**) cross section at different enlargements and (**c**) top view of the active layer.

Table 4. Experimental and calculated (pore size distribution) rejection rate of solutes (membrane with initial properties).

Solutes	*R* Exp (%)	*R* Calc (%)
L-phénylalanine	5	6
Vitamin B12	20	21
Lysozyme	65	66
L-Tyrosine	5	7

3.2. Hydraulic Performance Loss and Adsorption in the Pore

The hydraulic performances of the membranes decreased as filtration tests of neutral solutes were performed (except for Vitamin B12). At the same time, selectivity performances increased, indicating an increase in the steric effect limiting the mass transfer. As shown in Table 2 for the first series, the average pore radius (calculation relative to each filtrated molecule) decreased, step by step, until a minimum value. Four clogging/blocking mechanisms are classically reported in the literature: complete blocking (the filtrated molecule blocks the pore inlet), standard blocking (adsorption of the molecules in the pore), intermediate blocking (formation of a non-continuous layer, blocking partially the flow inlet), and a cake filtration. This last one often occurs in dead-end filtration. In our case, it can be ruled out owing to the tangential flow and associated shear stress. The intermediate blocking cannot explain both permeate flux decrease and selectivity increase. Indeed, to that end, non-continuous layers should preferentially grow, clogging the larger pore inlet only. The first mechanism, i.e., pore blocking, cannot be a plausible hypothesis, because during filtration of amino acids, the pores that clog should be the smallest, which would result in a decrease in rejection rate. So, the only plausible explanation is protein or amino acid adsorption in the larger pores, reducing the flow rate and increasing the selectivity. F. Wand et al. [36] observed that standard blocking occurred first during ultrafiltration of colloid–water solutions (dead-end filtration experiments). K. Katsoufidou et al. [37] observed a rapid irreversible membrane fouling during ultrafiltration of humic acid due to internal pore adsorption.

Figure 3 shows the hydraulic permeability of the membrane divided by the one obtained just before the first filtration test of lysozyme for series 1 and 3 (presented in Table 2). These results were compared with two test series performed in the same operating conditions with another membrane (series 4 and 5). The hydraulic permeability decreased as filtration tests were carried out (quasi linear behavior), indicating the same adsorption kinetics, regardless of the studied membrane and its past experiments. Assuming one size pore distribution, the hydraulic permeability was proportional to the fourth power of the pore radius (Poiseuille flow). In these conditions, the pore radius estimated by the filtration tests of a neutral solute should vary according to the fourth-root dependence of the number of tests.

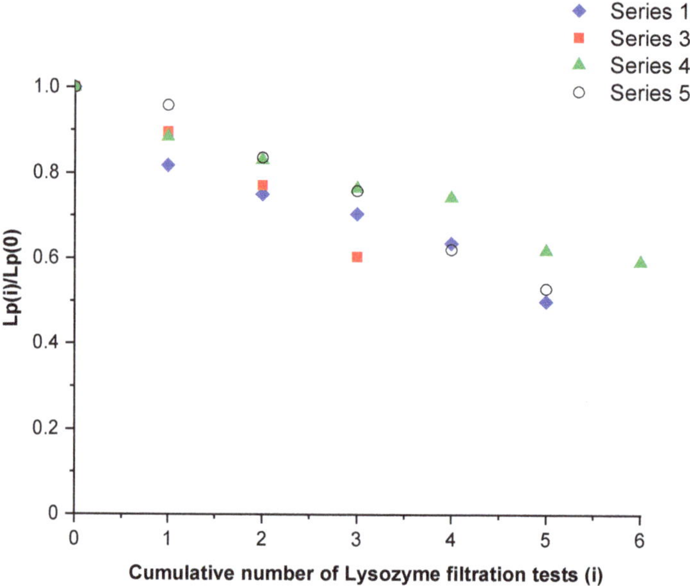

Figure 3. Relative hydraulic permeability versus the number of filtration tests of lysozyme performed.

Figure 4 compares the estimated mean pore radius for vitamin B12 and Lysozyme filtration tests (series 1) with the mean pore radius calculated using a linear function of the fourth-root of the cumulative number of lysozyme test. For the two cases, the estimated pore radius can be approximated by $r_p(i) = r_p(0) \times (1 - 0.165i)^{0.25}$.

The same investigation with pore size distribution was not possible owing to the number of adjustable parameters (i.e., $Np(i)$ and $r_p(i)$).

Steric hindrance after protein adsorption is required in order to predict the selectivity performances of the membrane. Classical models of adsorption or adsorption kinetics do not provide information about flow restriction in pores.

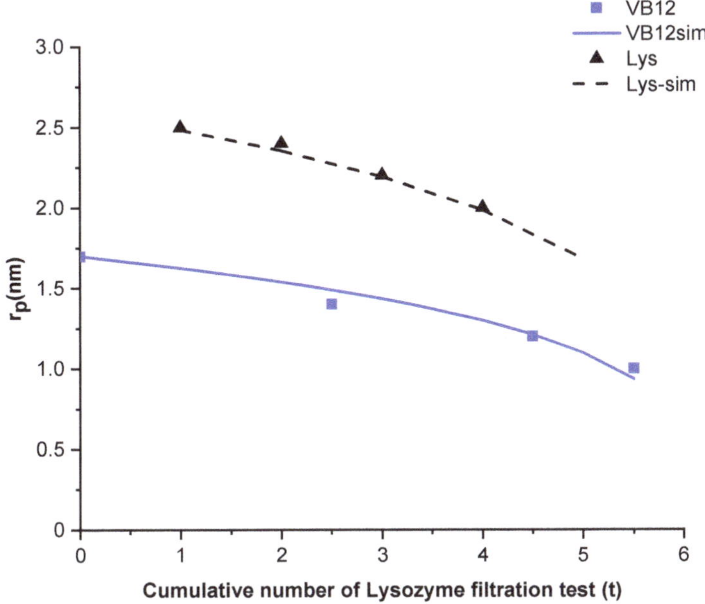

Figure 4. Mean pore radius estimated from filtration experiments and calculation of neutral solute (series 1: Vitamin B12 and Lysozyme).

In these conditions, to model these adsorption phenomena, we considered the adsorption of spherical molecules to be by uniform layers in the pore, reducing its radius of two stokes radius of the adsorbed molecule. This assumption is an arbitrary but required hypothesis for calculating the pore size distribution after molecule adsorption. The active layer thickness and the total number of pores were assumed to be unchanged (only their sizes can be modified). The calculation was carried out in all the pores with diameters twice larger than the molecule size. This procedure was numerically repeated until the calculated mass flow rate equaled the experimental permeation flux. When the equality of experimental and numerical hydraulic performances was reached, the model became predictive and capable of calculating the rejection rates of other filtrated molecules (i.e., Vitamin B12, *L*-Tyrosine, *L*-Phenylalanine, and Lysozyme).

Figure 5 shows the pore size distribution of the membrane after several filtration tests of lysozyme solution (series 1) compared to the initial one. This pore size distribution was obtained by equaling the experimental and calculated permeation fluxes. The results show that adsorption phenomena took place in the larger pores, partially clogging them and reducing their apparent diameters. Consequently, the pores with an initial size higher than 3.9 nm (corresponding to pores larger than adsorbed solute) completely disappeared and the pore quantity with a diameter in between the 1.8–2.8 nm range significantly increased.

To validate this modeling, rejection rates of the four filtrated molecules were calculated before and after lysozyme adsorption.

Experimental and calculated rejection rates of the studied molecules are given in Table 5 for the membrane before (initial properties) and after several filtration tests of lysozyme solutions. Taking into account that a pore size distribution provides a good description of membrane selectivity performances and its modification over time, from this new size distribution, the rejection rates of *L*-phenylalanine, *L*-Tyrosine, and Lysozyme increased and are in good agreement with the experimental results. Nevertheless, the rejection rate of vitamin B12 increased up to 60%, which is rather different regarding the experimental results (between 85% and 90%). These results are basically prior image and need to be further investigated and confirmed by other membranes and molecules.

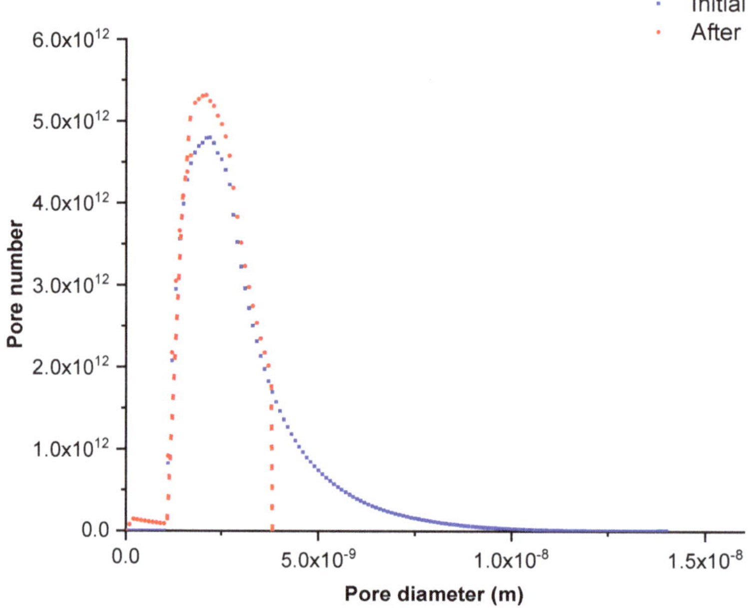

Figure 5. Initial and calculated pore size distribution after lysozyme filtration tests (series 1).

Table 5. Experimental and calculated rejection rates before and after lysozyme filtration.

Molecules	Initial Properties R/R_{calc}	After Lysozyme Filtration R/R_{calc}
L-phenylalanine	5%/6%	16%/13%
VB12	20%/21%	90%/60%
Lysozyme	65%/66%	100%/98%
L-tyrosine	5%/7%	17%/13%

4. Conclusions

In this work, filtration tests of four uncharged solutes were performed in order to investigate the steric exclusion effects. These tests were performed with new or regenerated membrane to investigate the pore size distribution and selectivity of the membrane regarding each solute. Taking into account a pore size distribution experimentally determined with nitrogen adsorption/desorption techniques made it possible to accurately describe the membrane selectivity performances. The simulated and experimental rejection rates of the four studied molecules are in good agreement for new membranes. Successive filtration tests showed that proteins and amino acids interact with the membrane (adsorption in the larger pore), reducing the hydraulic performances and increasing the selectivity. To describe this phenomenon, the pore size distribution was recalculated by considering molecule adsorption in the larger pore. In this way, the calculation was performed by equalizing experimental and numerical permeation flux. The simulated results provide a good perspective of trends in membrane selectivity for neutral solute filtration.

Author Contributions: Conceptualization, P.D. and A.P.; Methodology, S.M.M., P.D. and A.P.; Validation, S.M.M., P.D. and A.P.; Modelling P.D.; Investigation, S.M.M., P.D. and A.P.; Writing—Original Draft Preparation, S.M.M., P.D. and A.P.; Supervision, P.D. and A.P.

Funding: This research received no external funding

Conflicts of Interest: The authors declare no conflict of interest.

Glossary

C_p	Permeate concentration of solute (mol m^{-3})
C_r	Solute concentration in the feed solution (mol m^{-3})
D_∞	Diffusion coefficient of solute at infinite dilution (m^2 s^{-1})
J_v	Permeation flux (m^3 m^{-2} s^{-1})
K_c	Hindrance factor for convection (dimensionless) $K_c = 1 - 2.3\lambda + 1.154\lambda^2 + 0.224\lambda^3$
K_d	Hindrance factor for diffusion (dimensionless) $K_d = (2 - \varphi)\times(1 + 0.054\lambda - 0.988\lambda^2 - 0.441\lambda^3)$
L_p	Hydraulic permeability (m^3 m^{-2}memb)
$np(i)$	Pore number with a size $r_p(i)$
Q	Permeation flow rate (m^3 s^{-1})
$q(i)$	Permeation flow rate in pores with a size $r_p(i)$ (m^3 s^{-1})
r	Stokes radius of solute (m)
R	Observed rejection rate (dimensionless)
$Rcalc$	Calculated rejection rate
$R(i)$	Observed rejection rate of solute in pores with a size $r_p(i)$ (dimensionless)
r_p	Average pore radius (m)
$r_p(i)$	Pore radius
S_m	Membrane surface
t	Number of successive tests

Greek Letters

ΔP	Transmembrane pressure (Pa)
Δx	Active layer thickness (m)
λ	Radii ratio (dimensionless) $\lambda = r/r_p$
μ	dynamic viscosity of the filtrated solution (Pa s)
φ_i	steric partition coefficient relative to pores with a size $r_p(i)$ (dimensionless)

References

1. Baker, R.W. *Membrane Technology and Applications*, 3rd ed.; Wiley: Hoboken, NJ, USA, 2012.
2. Ho, W.S.W.; Sirkar, K.K. *Membrane Handbook*; Springer: New York, NY, USA, 1992.
3. Li, N.N.; Fane, A.G.; Ho, W.S.W.; Matsuura, T. *Advanced Membrane Technology and Applications*; Wiley: Hoboken, NJ, USA, 2008.
4. Pabby, A.K.; Rizvi, S.S.H.; Requena, A.M.S. *Handbook of Membrane Separations: Chemical, Pharmaceutical, Food and Biotechnological Applications*, 2nd ed.; CRC Press: Boca Raton, FL, USA, 2015.
5. Vandevivere, P.C.; Bianchi, R.; Verstraete, W. Review: Treatment and reuse of wastewater from the textile wet-processing industry: Review of emerging technology. *J. Chem. Technol. Biotechnol.* **1999**, *72*, 289–302. [CrossRef]
6. Arise, A.K.; Alashi, A.M.; Nwachukwu, I.D.; Ijabadeniyi, O.A.; Aluko, R.E.; Amonsou, E.O. Antioxidant activities of bambara groundnut (*Vigna subterranea*) protein hydrolysates and their membrane ultrafiltration fractions. *Food Funct.* **2016**, *5*, 2431–2437. [CrossRef] [PubMed]
7. Wang, X.; Li, C.; Du, C.; Gao, J.; Zhao, K.X.; Shi, R.; Jiang, Y. Plasma protein binding monitoring of therapeutic drugs in patients using single set of hollow fiber centrifugal ultrafiltration. *Bioanalysis* **2017**, *9*, 579–592. [CrossRef] [PubMed]
8. Schönherr, J. Water permeability of isolated cuticular membranes: The effect of pH and cations on diffusion, hydrodynamic permeability and size of polar pores in the cutin matrix. *Planta* **1976**, *128*, 113–126. [CrossRef]
9. Paganelli, C.V.; Solomon, A.K. The rate of exchange of tritiated water across the human red cell membrane. *J. Gen. Physiol.* **1957**, *41*, 259–277. [CrossRef]
10. Schaerp, J.; van der Bruggen, B.; Vandecasteerle, C.; Wilms, D. Influence of ion size and charge in nanofiltration. *Sep. Purif. Technol.* **1998**, *14*, 155–162. [CrossRef]

11. Spiegler, K.S.; Kedem, O. Thermodynamics of hyperfiltration (reverse osmosis): Criteria for efficient membrane. *Desalination* **1966**, *1*, 311–326. [CrossRef]

12. Ito, T.; Hioko, T.; Yamaguchi, T.; Shinbo, T.; Nakao, S.I.; Kimura, S. Development of a molecular recognition ion gating membrane and estimation of its pore size control. *J. Am. Chem. Soc.* **2002**, *124*, 7840–7846. [CrossRef]

13. Velicangil, O.; Howell, J.A. Estimation of the properties of high-flux ultrafiltration membranes. *J. Phys. Chem.* **1980**, *84*, 2991–2992. [CrossRef]

14. Bowen, W.R.; Mohammad, A.W.; Hilal, N. Characterisation of nanofiltration membranes for predictive purposes. Use of salts, uncharged solutes and atomic force microscopy. *J. Memb. Sci.* **1997**, *126*, 91–105. [CrossRef]

15. Déon, S.; Lam, B.; Fievet, P. Application of a new dynamic transport model to predict the evolution of performances throughout the nanofiltration of single salt solutions in concentration and diafiltration modes. *Water Res.* **2018**, *136*, 22–33. [CrossRef] [PubMed]

16. Dutournié, P.; Limousy, L.; Blel, W.; Déon, S.; Fievet, P. Understanding of ion transport in a Na-Mordenite membrane: Use of numerical modeling to estimate surface-solute interactions in the pore. *Ind. Eng. Chem. Res.* **2014**, *53*, 8221–8227. [CrossRef]

17. Bandini, S.; Morelli, V. Mass transfer in 1812 spiral wound modules: Experimental study in dextrose-water nanofiltration. *Sep. Pur. Technol.* **2018**, *199*, 84–96. [CrossRef]

18. Karlsson, M.; Ekeroth, J.; Elwing, H.; Carlsson, U. Reduction of irreversible protein adsorption on solid surfaces by protein engineering for increased stability. *J. Biol. Chem.* **2005**, *280*, 25558–25564. [CrossRef]

19. Norde, W. Driving forces for protein adsorption at solid surfaces. *Macromol. Symp.* **1996**, *103*, 5–18. [CrossRef]

20. Huang, T.; Anselme, K.; Sarrailh, S.; Ponche, A. High-performance liquid chromatography as a technique to determine protein adsorption onto hydrophilic/hydrophobic surfaces. *Int. J. Pharm.* **2016**, *497*, 54–61. [CrossRef]

21. Ponche, A.; Ploux, L.; Anselme, K. Protein/Material Interfaces: Investigation on Model Surfaces. *J. Adhes. Sci. Technol.* **2010**, *24*, 2141–2164. [CrossRef]

22. Robertson, B.C.; Zydney, A.L. Protein adsorption in asymmetric ultrafiltration membranes with highly constricted pores. *J. Colloid Interface Sci.* **1990**, *134*, 563–575. [CrossRef]

23. Comerton, A.M.; Andrews, R.C.; Bagley, D.M.; Yang, P. Membrane adsorption of endocrine disrupting compounds and pharmaceutically active compounds. *J. Membr. Sci.* **2007**, *303*, 267–277. [CrossRef]

24. Jepsen, K.L.; Bram, M.; Pedersen, S.; Yang, Z. Membrane fouling for produced water treatment: A review study from a process control perspective. *Water* **2018**, *10*, 847. [CrossRef]

25. Bu, F.; Gao, B.; Yue, Q.; Liu, C.; Wang, W.; Shen, X. The Combination of Coagulation and Adsorption for Controlling Ultra-Filtration Membrane Fouling in Water Treatment. *Water* **2019**, *11*, 90. [CrossRef]

26. Zhang, R.; Liu, Y.; He, M.; Su, Y.; Zhao, X.; Elimelech, M.; Jiang, Z. Antifouling membranes for sustainable water purification: Strategies and mechanisms. *Chem. Soc. Rev.* **2016**, *45*, 5888–5924. [CrossRef] [PubMed]

27. Tien, C.; Ramarao, B.V.; Yasarla, R. A blocking model of membrane filtration. *Chem. Eng. Sci.* **2014**, *111*, 421–431. [CrossRef]

28. Dutournié, P.; Limousy, L.; Anquetil, J.; Déon, S. Modification of the selectivity properties of tubular ceramic membranes after alkaline treatment. *Membranes* **2017**, *7*, 65. [CrossRef]

29. Bowen, W.R.; Welfoot, J.S.; Williams, P.M. Linearized transport model for nanofiltration: Development and assessment. *AIChE J.* **2002**, *48*, 760–773. [CrossRef]

30. Déon, S.; Dutournié, P.; Limousy, L.; Bourseau, P. The two-dimensional pore and polarization transport model to describe mixtures separation by nanofiltration: Model validation. *AIChE J.* **2011**, *57*, 985–995. [CrossRef]

31. Donnan, F.G. The theory of membrane equilibria. *Chem. Rev.* **1924**, *1*, 73–90. [CrossRef]

32. Déon, S.; Dutournié, P.; Bourseau, P. Modelling nanofiltration with Nernst-Planck approach and polarization layer. *AIChE J.* **2007**, *53*, 1952–1969. [CrossRef]

33. Bargeman, G.; Vollen broek, J.M.; Straatsma, J.; Schroën, C.G.P.H.; Boom, R.M. Nanofiltration of multi-component feeds. Interactions between neutral and charged components and their effect on retention. *J. Membr. Sci.* **2005**, *247*, 11–20. [CrossRef]

34. Bellona, C.; Drewes, J.E.; Xu, P.; Amy, G. Factors affecting the rejection of organic solutes during NF/RO treatment—A literature review. *Water Res.* **2004**, *38*, 2795–2809. [CrossRef]

35. Escoda, A.; Fievet, P.; Lakard, S.; Szymczyk, A.; Déon, S. Influence of salts on rejection of polyethyleneglycol by an NF organic membrane: Pore swelling and salting-out effects. *J. Membr. Sci.* **2010**, *347*, 174–182. [CrossRef]
36. Wang, F.; Tarabara, V.V. Pore blocking mechanisms during early stages of membrane fouling by colloids. *J. Colloid Interface Sci.* **2008**, *328*, 464–469. [CrossRef] [PubMed]
37. Katsoufidou, K.; Yiantsios, S.G.; Karabelas, A.J. A study of ultrafiltration membrane fouling bu humic acids and flux recovery by backwashing: Experiments and modelling. *J. Membr. Sci.* **2005**, *266*, 40–50. [CrossRef]

water

MDPI

Article

Use of Lignite as a Low-Cost Material for Cadmium and Copper Removal from Aqueous Solutions: Assessment of Adsorption Characteristics and Exploration of Involved Mechanisms

Salah Jellali [1,*], Ahmed Amine Azzaz [2], Mejdi Jeguirim [2], Helmi Hamdi [3] and Ammar Mlayah [4]

1   PEIE Research Chair for the Development of Industrial Estates and Free Zones, Center for Environmental Studies and Research (CESAR), Sultan Qaboos University, Al-Khoud, 123 Muscat, Oman
2   The institute of Materials Science of Mulhouse (IS2M), University of Haute Alsace, University of Strasbourg, CNRS, UMR 7361, F-68100 Mulhouse, France; amine.azzaz@uha.fr (A.A.A.); mejdi.jeguirim@uha.fr (M.J.)
3   Center for Sustainable Development, College of Arts and Sciences, Qatar University, P.O. Box 2713, Doha, Qatar; hhamdi@qu.edu.qa
4   Georesources laboratory, Water Research and Technologies Center (CERTE), Echo park of Borj Cedria, Carthage University, P.O. Box 273, Soliman 8020, Tunisia; ammarmlayah17@gmail.com
*   Correspondence: s.jellali@squ.edu.om

Abstract: Lignite, as an available and low-cost material, was tested for cadmium (Cd) and copper (Cu) removal from aqueous solutions under various static experimental conditions. Experimental results showed that the removal efficiency of both metals was improved by increasing their initial concentrations, adsorbent dosage and aqueous pH values. The adsorption kinetic was very rapid for Cd since about 78% of the totally adsorbed amounts were removed after a contact time of only 1 min. For Cd and Cu, the kinetic and isothermal data were well fitted with pseudo-second order and Freundlich models, respectively, which suggests that Cd/Cu removal by lignite occurs heterogeneously on multilayers surfaces. The maximum Langmuir's adsorption capacities of Cd and Cu were assessed to 38.0 and 21.4 mg g^{-1} and are relatively important compared to some other lignites and raw natural materials. Results of proximate, scanning electron microscopy/energy dispersive X-ray spectroscopy (SEM/EDS), Fourier transform infrared spectroscopy (FTIR) and X-Ray diffraction (XRD) showed that the removal of these metals occurs most likely through a combination of cation exchange and complexation with specific functional groups. The relatively high adsorption capacity of the used lignite promotes its future use as a low cost material for Cd and Cu removal from effluents, and possibly for other heavy metals or groups of pollutants.

Keywords: lignite; heavy metals; adsorption; batch; isotherm; mechanism

Citation: Jellali, S.; Azzaz, A.A.; Jeguirim, M.; Hamdi, H.; Mlayah, A. Use of Lignite as a Low-Cost Material for Cadmium and Copper Removal from Aqueous Solutions: Assessment of Adsorption Characteristics and Exploration of Involved Mechanisms. *Water* **2021**, *13*, 164. https://doi.org/10.3390/w13020164

Received: 13 December 2020
Accepted: 6 January 2021
Published: 12 January 2021

Publisher's Note: MDPI stays neutral with regard to jurisdictional claims in published maps and institutional affiliations.

1. Introduction

The contamination of water resources by heavy metals contained in industrial effluents is an important worldwide concern due to their toxicity, low biodegradability and high accumulation capacity in water-living organisms [1]. When transferred to humans through the food chain, these metals could be accumulated in various body organs and tissues causing life-threating illness and possible damages to vital systems [1]. Cadmium and copper are among the most toxic heavy metals. Indeed, exposure to cadmium, mainly used in batteries and the coating and plating industry, could result in stomach pains, bone fracture, possible infertility as well as serious damages to the central nervous and immune systems [2]. High ingested amounts of copper could increase infection frequencies, cardiovascular risks, and alterations in cholesterol metabolism [3]. The removal of these pollutants from industrial wastewater before discharge into the environment is, therefore, an urgent task to be appropriately achieved.

Many technologies have been developed for heavy metal removal from industrial effluents. They include chemical precipitation, reduction, electrodialysis, and membrane

Water **2021**, *13*, 164. https://doi.org/10.3390/w13020164   https://www.mdpi.com/journal/water

separation [4–7]. These technologies have shown interesting removal efficiencies at laboratory scale. However, their upscaling to real cases has been hindered by several limitations such as high capital and exploitation costs, very sensitive operational conditions, the use of large amounts of chemicals, and the production of secondary sludge that has to be sustainably handled [7].

The adsorption technique has been pointed out as a promising, eco-friendly and low-cost method for heavy metal removal from aqueous solutions [8–12]. In general, adsorbents with high pH, important specific surface area, well developed microporosity and that are rich in specific functional groups are used for the efficient removal of heavy metals form effluents. Once the adsorption process is completed, the loaded-adsorbents could be regenerated and heavy metals could be recovered for a further subsequent use with respect to the circular economy and sustainable development concepts [13,14]. Various mineral and organic materials have been tested for Cd(II) and Cu(II) removal from synthetic/real wastewaters. They include powdered marble wastes [15], clays [16], lignocellulosic biomasses [17], biochars [18,19], activated carbons [20,21] etc.

Lignite, often referred as brown coal, is a sedimentary rock that is naturally formed from naturally compressed peat. It is a low cost material and available in various countries at high amounts [22]. Besides its classical use for energy generation, raw lignite has been tested for the removal of various pollutants from aqueous solutions such as phenol and chlorophenol [23], trichloroethylene [24], phosphorus [25] and ammonium [26]. A special focus has been dedicated to the use of these raw materials as efficient adsorbents for heavy metals form aqueous solutions [27–30]. As such, these materials have relatively high contents of functional groups such as carboxyl, alcoholic and carbonyl groups that could complex with metal ions [27,31]. Furthermore, they have a large cation exchange capacity allowing cationic pollutants to be adsorbed [32].

Laboratory investigations regarding heavy metal removal by raw lignite have demonstrated that it could be considered as an interesting candidate for single metal removal form aqueous solutions [27,30,32]. However, when in multicomponent systems, the competition phenomenon decreases their maximal adsorption capacities. In both systems, results are sometimes contradictory and highlight that the ability of heavy metal retention depends on their physico-chemical characteristics as well as the lignite properties [33,34]. For instance, Pehlivan et al. [34] studied the adsorption of Pb, Cd, Cu, Ni, and Zn on several Turkish lignite materials. They showed that the order of metal adsorption ability depends on the lignite type. However, Pb and Cu were the most adsorbed components in contrast to Ni and Zn. Furthermore, Pentari et al. [30] studied the removal of Pb, Cd, Zn, and Cu from aqueous solutions by a lignite from Greece. They found that in single mode, the ability adsorption order was as follows: Pb > Cd > Cu > Zn. In a multicomponent system, they showed that Cu sorption was the most affected, while Zn adsorption capacity remained quasi-constant. For the same metals in a multicomponent system, Doskočil and Pekař [33] found a different order: Pb > Cu > Zn > Cd. On the other hand, the involved mechanisms during heavy metal adsorption onto raw lignite are still not well identified. An in-depth kinetic and isothermal modeling study combined with lignite analysis before and after metal adsorption by using advanced techniques could reduce this gap.

As a consequence, the principal objectives of this work were: (i) to assess the adsorption characteristics of Cd and Cu onto a Tunisian lignite under various experimental conditions including contact time, initial concentration, pH, adsorbent dosage, and competition with other metals, (ii) to compare the efficiency of this lignite with raw and modified lignite available in the literature, (iii) to better understand the involved adsorption mechanisms of Cd and Cu through the combination of an in-depth modeling study and various physico-chemical analyses.

2. Materials and Methods

2.1. Adsorbent Preparation and Characterization

Raw lignite was collected from the Cap Bon region (northeastern part of Tunisia). This lignite was used in its natural state after a drying step at 60 °C for 24 h followed by manual grinding in ceramic grinder. The fraction with diameter size lower than 63 μm was selected for the adsorption tests. The preliminary characterization of the used lignite included the determination of its: (i) mineral composition by X-ray fluorescence spectrophotometer (Philips, Eindhoven, The Netherlands), (ii) Brunauer–Emmett–Teller BET specific surface area through N_2 gas adsorption method using a gas adsorption analyzer (Quantachrom Autosorb 1 sorptiometr), and (iii) pH of zero-point-charge value (pH_{ZPC}) [35].

Furthermore, advanced analyses of the lignite before and after metals adsorption were performed for a better identification of the involved mechanisms. They included the assessment of: (i) the morphology and qualitative composition through scanning electron microscopy (SEM) coupled with energy dispersive X-ray (EDX) (Philips model FEI model Quanta 400 apparatus, Amsterdam, The Netherlands), (ii) proximate analysis using a TGA/DSC3+ device (Mettler-Toledo, Greifense, Switzerland), and (iii) the existing functional groups through a Fourier transform infrared (FTIR) analysis using an Equinox 55 spectrometer (Brucker, Billerica, MA, USA). The FTIR spectra were assessed between 4000 and 400 cm^{-1} with a resolution of 1 cm^{-1}. Experimental protocols for the above-cited analyses have already been detailed in a previous paper [36].

2.2. Synthetic Heavy Metals Solutions Preparation and Analysis

Cadmium nitrate ($Cd(NO_3)_2$), copper nitrate ($Cu(NO_3)_2$), lead nitrate ($Pb(NO_3)_2$), and zinc nitrate ($Zn(NO_3)_2$) were used for the preparation of four stock solutions at a concentration of 1 g L^{-1} each (Fisher Scientific, Waltham, MA, USA). These solutions were used throughout this study for the preparation of adsorption solutions at precise concentrations. Metal concentrations were measured thorough an atomic absorption spectrometer (AAS) with an air-acetylene flame (Perkin Elmer Analyst 200, Waltham, MA, USA). The wavelengths used for the analysis of the Cd, Cu, Pb and Zn were 228.8, 324.8, 283.3, and 213.9 nm, respectively. The initial pH values of the solutions were adjusted by using dilute sodium hydroxide or nitric acid.

2.3. Batch Adsorption Investigation

Cd and Cu removal efficiency from aqueous solutions by raw lignite was carried under static conditions (batch mode). It consists in shaking, at room temperature (20 ± 2 °C), a given mass of lignite in 50 mL of aqueous solution containing the metal at a fixed concentration for a desired contact time at 400 rpm by using a Variomag-poly15 magnetic stirrer. Then, the suspension was filtrated through 0.45 μm cellulose acetate filter before analysis with AAS. During this study, the effect of the following experimental conditions on Cd and Cu removal efficiency were assessed: (i) particle size distribution for four lignite granulometries: <63 μm; between 63 and 500 μm; 500–1000 μm, and 1000–2000 μm; (ii) the contact time of 1; 5; 10; 20; 40; 60, and 90 min; (iii) the initial aqueous pH for 2.0; 3.0; 4.0 and 5.0, and (iv) lignite dosages for 0.4; 1.0; 1.6; 2.0; 2.4; 3; 3.5 and 4.0 g L^{-1}. During these assays, the default following parameters were used: a lignite size fraction lower than 63 μm, a contact time of 90 min, an initial Cd or Cu concentration of 100 mg L^{-1}, an initial pH of 5 and a lignite dosage equal to 2 g L^{-1}. Finally, the competing effect was determined for a multicomponent solution containing Cd, Cu, Pb and Zn at constant concentrations of 30 mg L^{-1}. All these experiments were carried out in triplicate and mean values are reported in this work. The standard deviation for all assays was lower than 5%.

The adsorbed metal amount at a given moment 't', (qt) and the related removal yield (Yt) were determined as follows [37]:

$$q_t = \frac{(C_0 - C_t)}{D} \tag{1}$$

$$Y_e(\%) = \frac{(C_0 - C_t)}{C_0} \times 100 \tag{2}$$

where C_0 and C_t (mg L^{-1}) are initial metal concentrations and at a given time 't', respectively, and D is the used adsorbent dose (g L^{-1}).

Cadmium and copper adsorption kinetics were fitted to three standard models namely, pseudo-first order (PFO), pseudo-second order (PSO), and intraparticle and film diffusion. Moreover, the experimental isothermal data were fitted to Langmuir, Freundlich and Dubinin–Radushkevich (D-R) isotherm models. The agreement between the experimental and theoretical adsorbed amounts was assessed through the determination of the average percentage errors (*APE kinetic* and *APE isotherm*) as follows:

$$APE_{kinetic}(\%) = \frac{\sum \left| (q_{t,exp} - q_{t,theo})/q_{t,exp} \right|}{N} \times 100 \tag{3}$$

$$APE_{isotherm}(\%) = \frac{\sum \left| (q_{e,exp} - q_{e,theo})/q_{e,exp} \right|}{N} \times 100 \tag{4}$$

where $q_{t,exp}$ and $q_{t,theo}$ (mg g^{-1}) are the experimental and the theoretical adsorbed amounts at a given time 't'; $q_{e,exp}$ and $q_{e,theo}$ (mg g^{-1}) are the experimental and the theoretical adsorbed masses at equilibrium.

3. Results

3.1. Lignite Characterization

The XRF analysis of the lignite showed that it contains various minerals. Silica, Sulfur, iron, and aluminum exist at relatively high contents of 6.2%, 5.1%, 2.6%, and 2.5% (dry mass), respectively. Lower contents of 0.06%, 0.46%, 1.21%, and 0.21% were observed for Na, K, Ca, and Mg, respectively. Some of these elements could be exchanged with Pb(II) during its adsorption by lignite. On the other hand, the used material has higher concentrations of Si, Al, Fe and Na and lower Ca and Mg content than a commercial and natural lignite from Czech Republic [38,39]. The BET surface area of the lignite was assessed to 11.2 m^2 g^{-1}, which is in the range of BET values reported for Greek lignites (between 3.6 and 23.8 m^2 g^{-1}) [30]. For instance, it is about 12-fold higher than a lignite from Poland (0.91 m^2 g^{-1}) [23], and about two-fold higher than a Hungarian lignite (5.3 m^2 g^{-1}) [25]. This relatively higher surface area suggests that the used lignite could exhibit more metal adsorption capacities compared to other lignites reported in the literature. On the other hand, its pH$_{ZPC}$ was estimated to 3.6. It is lower than that of a Czech lignite (5.0) [38], and a Polish lignite (6.2) [23]. In contrast, it is higher than pH$_{ZPC}$ of a Hungarian lignite (2.6) [25]. Since the net charge surface of the adsorbent becomes negative for aqueous pH values higher than pH$_{ZPC}$, the removal of positively charged metals by the current lignite through electrostatic attraction will be favored for a wide pH range.

3.2. Batch Adsorption Results

3.2.1. Effect of Lignite Granulometry

The impact of lignite granulometry on Cd and Cu removal from the synthetic solution was carried out for initial metal concentrations of 100 mg L^{-1} each, initial pH of 5 and a dosage of 2 g L^{-1}. The experimental results (Figure 1) showed that coarser is the lignite fraction, lower is Cd or Cu removal efficiency. Accordingly, the highest adsorbed amounts of Cd (26.1 mg g^{-1}) and Cu (18.6 mg g^{-1}) were observed for the finest granulometry (<63 µm). These adsorbed Cd and Cu amounts decreased by about 25% and more than 65%, respectively when lignite particles of 1–2 mm size were used. This outcome could be attributed to the lower microporosity and surface area generally observed for coarser media [40]. Similar behavior was observed by PN and CP [41] when investigating an oily effluent treatment by hard wood based adsorbents. They reported a decrease of the removal efficiency by about 30% when the average size particles were increased from 0.8 to 3.5 mm.

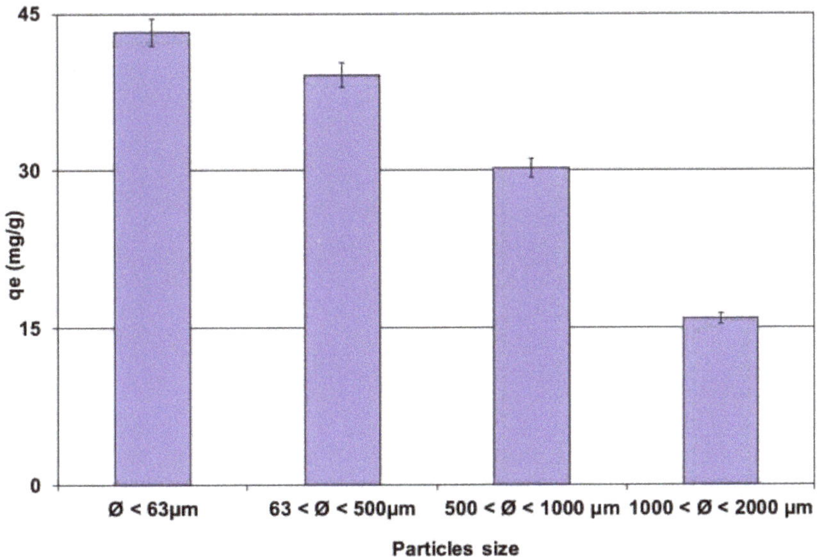

Figure 1. Impact of lignite particle size on Cd and Cu removal efficiency (C_0 = 100 mg L^{-1}, D = 2 g L^{-1}; t = 60 min; pH = 5; T = 20 ± 2 °C).

3.2.2. Effect of Contact Time—Kinetic Study

The removal of Cd and Cu by the used lignite is clearly a time-dependent process as shown in Figure 2. Indeed, their removal was very rapid at the beginning of the adsorption (especially for Cd) since about 78% and 44% of the totally removed amounts were adsorbed after only 1 min for Cd and Cu, respectively. This finding suggests that Cd removal occurs mainly through surface reactions [11]. The high reactivity of the used lignite could result in an important energy saving when such process is scaled up for field investigations. After this contact period, the adsorbed amounts continue to increase but at much slower rate. This behavior is linked to an intraparticle diffusion inside the pores of the lignite and adsorption by functional groups through complexation process [42]. The equilibrium state which corresponds to quasi-constant adsorbed amounts was reached after approximately 60 min for both metals. This time is 6-fold lower than the one reported by Havelcova et al. [38] when investigating Cd, Cu and Zn removal by a local lignite from south Moravian coal field (Czech republic) and 12-fold lower than the duration observed by Pentari et al. [43] for Cd removal by a raw and iron doped Greek lignite. However, relatively similar equilibrium contact times were observed for Cd and Cu removal by various Greek lignites [30]; Cu, Pb and Ni removal by two Turkish lignites [32]; Zn removal by a commercial lignite from South Korea [44] and Cr(VI) adsorption onto a raw Iranian lignite [28]. For economic reasons, low contact times, ensuring percentages removal of more than 80% of the totally adsorbed Cd or Cu amounts could be used in real case application. These durations correspond to only 5 and 20 min for Cd and Cu, respectively.

At equilibrium, the lignite sample used in this study removed Cd better than Cu. Indeed, the adsorbed Cd amount by was assessed to 26.1 mg g^{-1} which is about 41% higher than Cu (Figure 2). A Similar trend was observed by various studies dealing with heavy metal adsorption onto lignite [30,38]. This behavior is mainly imputed to their physico-chemical properties, especially electronegativity, ionic potential and ionic radius [45].

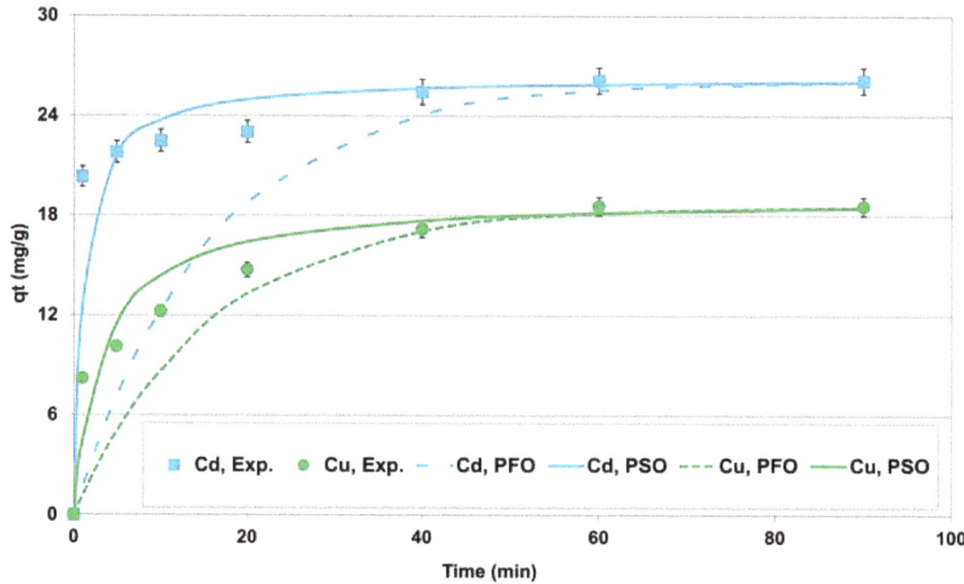

Figure 2. Kinetics of Cd and Cu removal by lignite and its fitting with PFO and PSO models (C_0 = 100 mg L^{-1}, D = 2 g L^{-1}; t = 60 min; pH = 5; T = 20 $\pm$ 2 °C; Exp.: Experimental value).

The parameters of the three theoretical models: PFO, PSO and diffusion models are given in Table 1. Based on these calculated parameters, it can be clearly deduced that the PFO model does not appropriately fit to the experimental data. In fact, the corresponding determination coefficients were low: 0.553 and 0.883 for Cd and Cu, respectively. The calculated APE between the measured and calculated adsorbed masses were high since they are about 28% for Cd and 22% for Cu. Figure 2 confirms this finding since the dashed lines are far from the measured data.

Table 1. Kinetic parameters of Cd and Cu removal by the used lignite (C_0 = 100 mg L^{-1}, D = 2 g L^{-1}; t = 60 min; pH = 5; T = 20 $\pm$ 2 °C).

	Cadmium	Copper
$q_{e,exp}$ (mg g^{-1})	26.1	18.6
Pseudo-first-order-model (PFO)		
$k1$ (min^{-1})	0.063	0.063
R^2	0.533	0.883
APE (%)	28.9	22.5
Pseudo-second-order-model (PSO)		
$q_{e,theo}$ (mg g^{-1})	26.4	19.2
K_2 (g mg^{-1} min^{-1})	0.034	0.016
R^2	0.894	0.932
APE (%)	6.9	11.7
Diffusion model		
D_f ($\times 10^{-14}$ m^2 s^{-1})	4.68	2.04
R^2	0.750	0.844
D_{ip} ($\times 10^{-14}$ m^2 s^{-1})	2.26	2.16
R^2	0.936	0.999

In contrast, the PSO model fits well the experimental data for both metals. The related determination coefficients (>0.89) were much higher than those for the PFO model. Furthermore, the APE for Cd and Cu were determined to only 7% and 12% for Cd and Cu, respectively and therefore were lower than those obtained for the PSO model. The theoretical adsorbed amounts of Cd and Cu at equilibrium ($q_{e,theo}$), were very close to the experimental ones (Figure 2 and Table 1), with different percentages of about 1% and 3%, respectively. Therefore, under the used experimental conditions, the PSO model is more suitable in fitting the Cd and Cu removal onto lignite. This model suggests that the rate limiting step might be chemical adsorption involving valency forces through sharing or exchange of electrons between these two metals and lignite [46].

The analysis of the adsorption of Cd and Cu onto the used raw lignite through the application of film and intraparticle diffusion models indicated clearly that that the adsorption process proceeds by surface interactions at earlier stages (short times, less than 5 min) and by intraparticle diffusion at later stages (Figure 2). For Cd, the film diffusion coefficient (through boundary layer) is about two times higher than the one of the intraparticle diffusion. This explains the very high adsorption percentage observed after only 1 min (Figure 2). This finding confirms that intraparticle diffusion process controls significantly the rate of Cd adsorption onto the lignite. Similar observations were reported by Pehlivan et al. [32] when studying Pb, Cu, Ni, and Zn removal by Turkish lignites. For Cu, the film and intraparticle diffusion coefficients are quite similar (Table 1) indicating that both processes (boundary layer and intraparticle diffusion) control its removal by the raw lignite.

3.2.3. Effect of Initial Aqueous pH

The effect of the initial aqueous pH on Cd and Cu removal efficiency was performed under the experimental conditions presented in Section 2.3. The experimental results showed that for both metals, the removed amounts increased with the increase of the aqueous pH (Figure 3). As such, for an initial pH of 2, the adsorbed amounts of Cd and Cu were 10.9 and 5.4 mg g^{-1}, respectively. At a pH of 5, these quantities reached 26.1 and 18.6 mg g^{-1} which are about 2.4 and 3.4 times higher than those registered at an initial pH of 2. This outcome could be explained by the lignite surface charge that is positive at aqueous pH lower than the pH$_{ZPC}$ (3.6), which will repulse the two cationic metals. Furthermore, at this pH range, abundant H$^+$ ions in the solution will compete with Cd and Cu over the adsorption sites. However, when the used aqueous pH value is higher than the pH$_{ZPC}$, the lignite surface will carry more negative charges and will consequently favor Cd and Cu adsorption through electrostatic reactions. At this pH range, H$^+$ and other exchangeable cations such as Ca^{2+}, Mg^{2+}, Na$^+$, and K$^+$ could be exchanged with these metals and released in the aqueous solutions. Similar trends were reported by several studies dealing with heavy metals or cationic pollutant adsorption onto raw or modified lignites [32,38] and other adsorbents [35,42,47,48]. For instance, Pehlivan et al. [32] showed that increasing pH form 2 to 6 increased Cu removal efficiency yield by a raw Turkish lignite from about 8% to more than 88%.

3.2.4. Effect of Lignite Dosage

The impact of lignite dose on Cd and Cu removal from aqueous solutions at an initial concentration of 100 mg L^{-1}, an initial pH of 5 and 60 min of contact time is given in Figure 4. Their removal yields increase with the increase of the lignite dose. As such, rising the dose from 0.4 to 3 g L^{-1}, Cd and Cu removal efficiencies increased from about 40.5% and 19.8% to 61.9% and 46.7%, respectively. This important increase is linked to the presence of more available adsorption sites that could interact with Cd or Cu when using higher doses. Starting from a used dose of 3.5 g L^{-1}, Cd or Cu removal efficiency remains approximately constant due to the saturation of lignite particles. Such high removal efficiencies observed for relatively low lignite doses is a real asset for larger applications at industrial levels.

A similar trend was reported by Binabaj and Ramezanian [28] and Gurses et al. [49] when studying chromium(VI) and methylene blue removal by raw lignites, respectively.

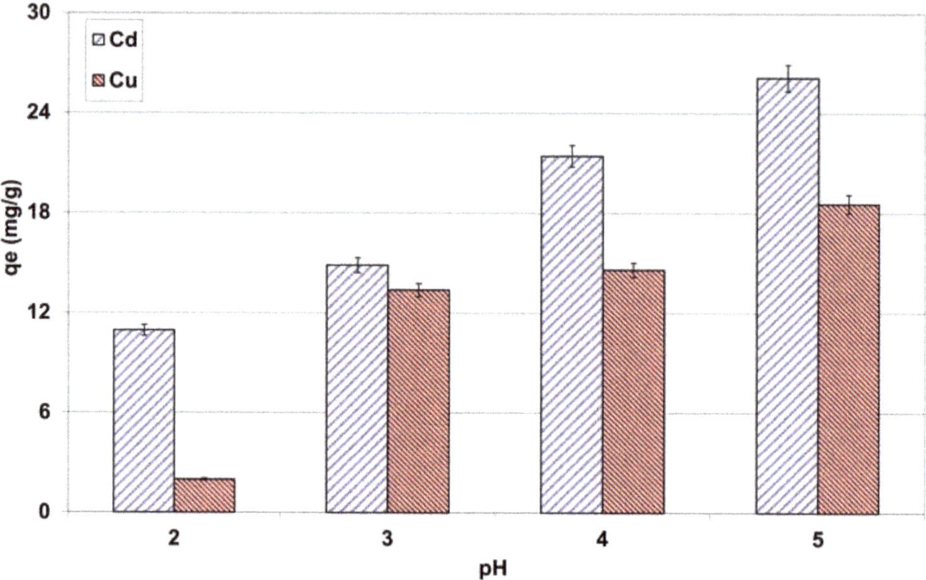

Figure 3. Impact of initial aqueous pH on Cd and Cu removal efficiency by lignite (C_0 = 100 mg L^{-1}, D = 2 g L^{-1}; t = 60 min; T = 20 ± 2 °C).

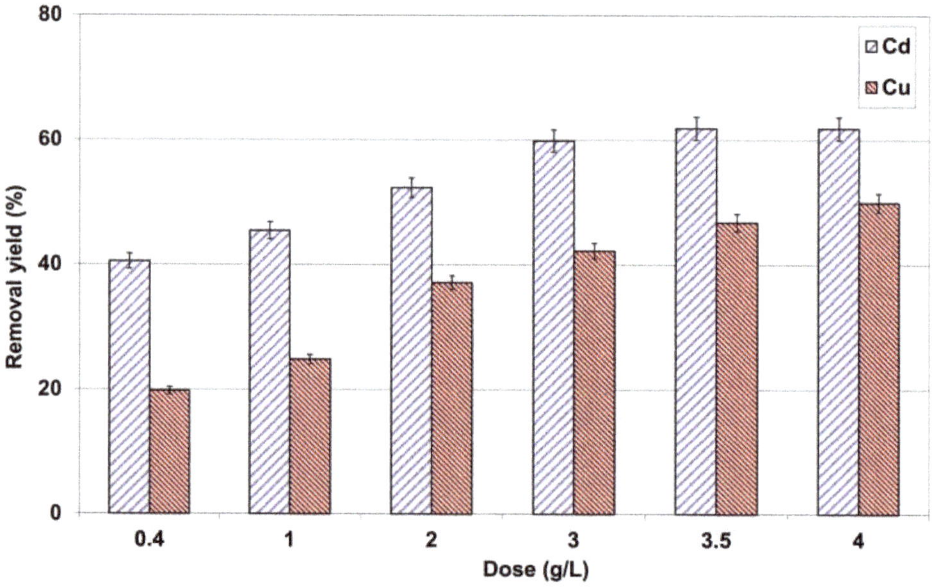

Figure 4. Impact of lignite dose on Cd and Cu removal yields (C_0 = 100 mg L^{-1}, t = 60 min; pH = 5; T = 20 ± 2 °C).

3.2.5. Competing Effect

Industrial effluents generally contain a complex mixture of organic and non-organic pollutants. The effect of the presence of Pb and Zn on the adsorption of Cd and Cu by the raw lignite was performed according to the experimental conditions described in Section 2.3. The experimental results (Figure 5) showed that for single mode, as reported by Pentari et al. [30], the adsorption efficiency of the studied metals was as follows: Pb > Cd > Cu > Zn. The corresponding removal percentages were estimated at 100%, 74.6%, 71.7% and 44.8%, respectively. Similar observations were also reported by Pehlivan et al. [34] who studied the adsorption of Pb, Cd, Cu, Ni, and Zn on several Turkish lignites. They showed that for the majority of the studied lignites, Pb, Cu and Cd were the most adsorbed metals whereas Ni and Zn were the less adsorbed ones.

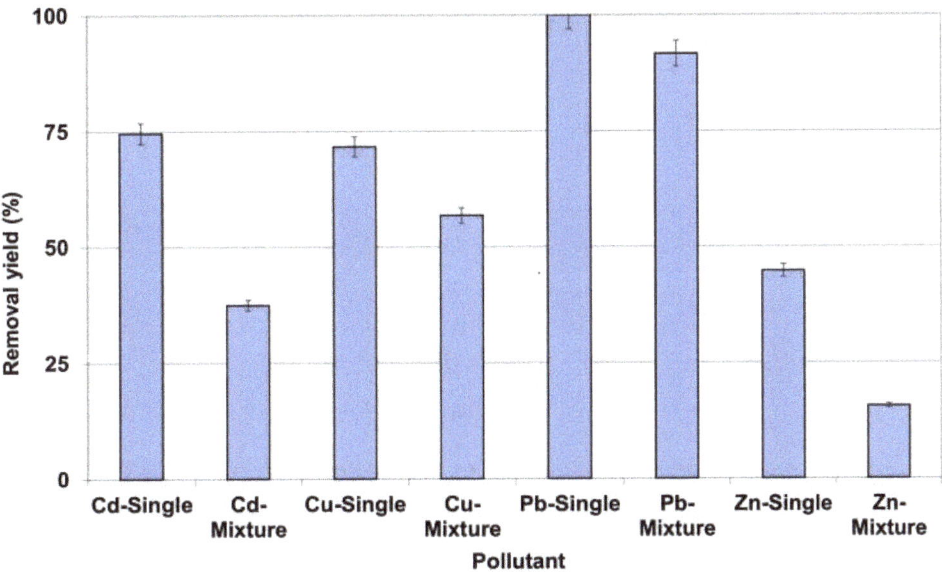

Figure 5. Cd, Cu, Pb and Zn removal yield in single and multicomponent systems ($C_0 = 30$ mg L^{-1} for all metals; D = 2 g L^{-1}; t = 60 min; pH = 5; T = 20 ± 2 °C).

In the multicomponent system, all the removal efficiencies of the four heavy metals significantly decreased due to competition. Accordingly, the highest removal decrease was observed for Cd (37.1%) and Zn (29.2%) and the adsorption pattern shifted to: Pb > Cu > Cd > Zn with removal efficiencies of 91.7%, 56.8%, 37.5% and 15.6%, respectively. Consequently, when treating real wastewater containing a mixture of heavy metals, Pb and Cu will be favorably retained by the current lignite compared to other metals. Similar trends were reported by Allen and Brown [50] when investigating the removal of Cd, Cu and Zn by an Irish lignite in a single and multicomponent systems. Their experimental results showed that in single metal sorption mode, the adsorption ability decreases in the order Cd > Cu > Zn. However, in the multicomponent assays, this order changed to: Cu > Cd > Zn. This result is in a contradiction with the one reported by Pentari et al. [30] who pointed out that in multicomponent system containing Pb, Cd, Cu, and Zn, copper adsorption was the most significantly influenced by the presence of the other elements with a total yield reduction of about 5%. This behavior should not be only attributed to metal characteristics including size, electronegativity, availability and hydration energy [15] but also to the physico-chemical properties of the lignite such as the donor atoms abundance (oxygen, nitrogen, sulphur [51]).

3.2.6. Isotherm Adsorption

For the experimental conditions cited in Section 2.3, the data of Cd and Cu adsorption isotherms in comparison with the predicted ones using Freundlich, Langmuir and D-R models are given in Figure 6. Table 2 gives the parameters of these models as well as their fitness to the experimental data.

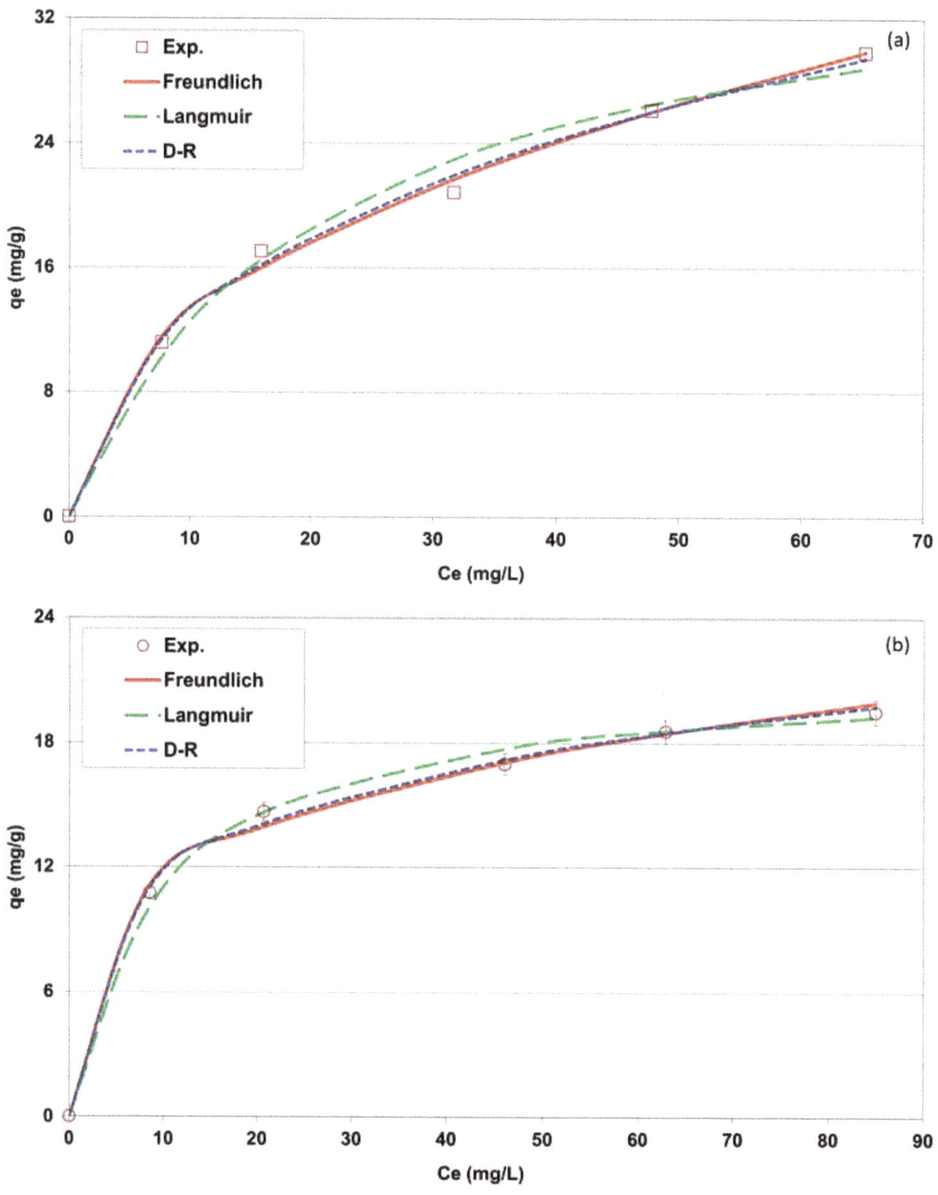

Figure 6. Isotherm experimental and fitted data with Freundlich, Langmuir and Dubinin–Radushkevich (D-R) models for Cd (**a**) and Cu (**b**) removal by lignite (D = 2 g L^{-1}; t = 60 min; pH = 5; T = 20 ± 2 °C; Exp.: Experimental value).

Table 2. Adsorption isotherm parameters of Cd and Cu removal by lignite (D = 2 g L^{-1}; t = 60 min; pH = 5; T = 20 ± 2 °C).

Metal	Cd	Cu
Freundlich model		
Freundlich constant: K_F	4.7	6.4
Freundlich constant: n	2.3	3.9
Determination coefficient: R^2	0.991	0.983
Average percentage error: APE (%)	2.8	2.3
Langmuir model		
Langmuir's maximum adsorption capacity; q_m (mg g^{-1})	38.0	21.4
Langmuir constant: K_L	0.048	0.104
Determination coefficient: R^2	0.979	0.957
APE (%)	5.4	2.5
D-R model		
DR's maximum adsorption capacity; q_m (mg g^{-1})	124.8	40.5
E (kJ mol^{-1})	10.7	13.4
Determination coefficient: R^2	0.990	0.990
APE (%)	2.3	1.5

It can be clearly demonstrated from Figure 6 and Table 2 that the three applied models fit very well with the experimental data. Nevertheless, the highest determination coefficients (R^2) (0.990 for Cd and for Cu) and the lowest APE (2.3% for Cd and 1.5% for Cu) were observed for D-R model. This model predicts, however, very high adsorption capacities for both Cd and Cu with respective values of 124.8 and 40.5 mg g^{-1}, respectively. These values are unrealistic considering the isotherm curves shape (Figure 6) and are mainly imputed to this model's used assumptions especially the one related to the uniformity and homogeneity of the adsorbent's microporous structure [35]. The calculated free energy (E = 1/$\sqrt{2\beta}$) was assessed to 10.7 and 13.4 kJ mol^{-1} for Cd and Cu, respectively. Both of them are higher than 8 kJ mol^{-1}, indicating that Cd and Cu removal by the used lignite was mainly chemical [49]. It is worth mentioning that compared to Langmuir model, The Freundlich one fits better the experimental data with higher R^2 coefficients and lower APE values (Table 2). This outcome indicates that both Cd and Cu adsorption by the used lignite occurs heterogeneously on multilayer surfaces through chemical processes [28].

On the other hand, the Freundlich constant 'n' values were 2.3 and 3.9 for Cd and Cu, respectively. They are in the range of 1–10, which indicates that the adsorption of these two heavy metals by the used lignite is a favorable process. Values in the same range were determined by Pentari et al. [30] when studying Cd and Cu removal by a raw Greek lignite. Moreover, the highest Langmuir's parameter values (obtained for the lowest used initial concentrations) '$R_L = \frac{1}{1+K_{L*}C_0}$' were estimated to 0.41 and 0.24 for Cd and Cu, respectively. They are lower than 1, suggesting that lignite could be considered as a favorable material for Cd and Cu retention from aqueous solutions.

The Langmuir's maximum adsorption capacities of Cd and Cu were determined to 38.0 and 21.4 mg g^{-1}, respectively. A comparison of the used lignite efficiency in removing Cd and Cu with other lignites (Table 3) clearly shows that it can be considered as a promising medium for the removal of heavy metals form wastewater. As a matter of fact, its Cd adsorption capacity was about 5.6, 1.7 and 1.5 times higher than a lignite activated carbon from China [52], a HNO$_3$ treated lignite activated carbon [52], and a raw lignite form Greece [43]. Its Cu adsorption capacity was higher than a Beypazari lignite [51], Ilgin lignite and Beysehir lignite [32]. However, other tested lignites exhibited higher adsorption capacities [27,30,53]. It is important to underline that the observed lignite adsorption capacities of Cd and Cu are also higher than various other agricultural, animal and industrial wastes [54,55]. As compared to natural agricultural wastes, Cd adsorption capacity of the studied lignite was about 7.5 and 2.3 higher than corncob [56] and rice husk [57], respectively. Moreover, Cu removal was 2.8 and 2.1 times more important than Banana peels [58] and grape stalks [59], respectively.

Table 3. Comparison of the used lignite removal efficiency of Cd and Cu with other lignites from various regions.

Material	BET Surface Area $(m^2\,g^{-1})$	Adsorption Conditions	Cd Langmuir Adsorption Capacity $(mg\,g^{-1})$	Cu Langmuir Adsorption Capacity $(mg\,g^{-1})$	Reference
Lignite, Beypazarı, Turkey	2.56	$C_0 = 0.00015$–0.0015 M; pH = 4; D = 10 g L^{-1}; t = 1 h ; T = 20 °C	-	1.62	[51]
Ilgin lignite, Konya, Turkey	2.06	$C_0 = 0.0025$–0.025 M; pH = 4.5; D = 4 g L^{-1}; t = 120 min; T = 20 °C	-	17.8	[32]
Beysehir lignite, Konya, Turkey	2.96		-	18.9	
Tyul'gansk lignite, South Ural Basin Russia	-	$C_0 = 0.1$ M; pH = 4.1; D = 40 g L^{-1}; t = 5 days; T = 25 °C	-	27.3	[27]
Tisul'sk lignite, Kansko-Achinsk Basin, Russia	-		-	27.3	
Lignite, South Moravian deposit, Mikulcice, Czech Republic	-	$C_0 = 0.001$–0.2 M; pH = 4–5; D = 20 g L^{-1}; t = 24 h; T = 25 °C	-	82.0	[53]
Lignite activated carbon (LAC), China	158.11	$C_0 = 0.00015$–0.0015 M; pH = 7.2; D = 1 g L^{-1}; t = 24 h ; T = 25 °C	6.8	-	[52]
Lignite activated carbon, treated with sodium dodecyl benzene sulfonate (SAC)	118.2		26.6	-	
HNO$_3$-treated lignite activated carbon (NAC)	185.07		22.8	-	
HNO$_3$-treated lignite activated carbon, post treated with sodium dodecyl benzene sulfonate (NSAC)	131.4		44.2	-	
Lignite, Drama, northern Greece	154	$C_0 = 0.00015$–0.015 M; pH = 4.5; D = 10 g L^{-1}; t = 12 h; T = 25 °C	25.5	-	[43]
Same lignite doped with Nano-scale-zero-valent-iron particles	109		34.7	-	
Lignite, Macedonia and Thrace, Greece	4.2	$C_0 = 0.00015$–0.015 M; pH = 4–5; D = 10 g L^{-1}; t = 45 min; T = 25 °C	51.5	42.7	[30]
Lignite, Cap Bon, Tunisia	11.2	$C_0 = 0.00047$–0.00197 M; D = 2 g L^{-1}; pH = 5; T = 20 ± 2 °C	38.0	21.0	Current study

As illustrated in Table 3, Cd or Cu adsorption efficiency by lignites depend not only on their specific surface area (case of the activated lignites [52]) but also on their functional group types and richness and the experimental conditions. The same trend was reported by Puglla et al. [19] who found relatively low adsorption capacities of Cd onto biochars derived from peanut shell (1.038 mg g^{-1}), chonta pulp (0.655 mg g^{-1}) and corncob (0.857 mg g^{-1}) even if they have interesting textural properties. It is important to underline that the structural and textural properties of the lignite could be improved through specific physical/chemical/thermal modifications [52,60,61]. For instance, Sun et al. [52] found that the chemical/thermal modification of lignite by a HNO$_3$ solution simultaneously increased its surface area (from 158.1 to 185.1 m^2 g^{-1}) and its surface carboxylic and lactone functional groups (from 0.55 to 1.26 mmol g^{-1}). Consequently, such modification increased Cd removal capacity from 6.8 to 26.6 mg g^{-1}.

3.3. Raw and Metal-Loaded Lignite Characterization and Adsorption Mechanism Exploration

Various techniques were used for the characterization of lignite before and after metal adsorption in order to get a better understanding of the probably involved mechanisms. Morphological properties of the lignite before and after adsorption of metals were investigated through imagery comparison using scanning electron microscopy and energy-dispersive X-ray spectroscopy (Figures 7 and 8).

Figure 7. SEM images of: (**a**) raw lignite; (**b**) after adsorption of cadmium; (**c**) after adsorption of copper (magnitude: ×2500).

It appears that the raw lignite (particle size lower than 63 μm) presents an irregular and heterogeneous surface. Moreover, similarly to the majority of natural carbonaceous materials, the used lignite does not present a concrete porosity (Figure 7a), which explain its measured specific surface area. SEM imaging reveals also the presence of crystalline structures on the surface. According to Zhang and Chen [62], raw uncrushed lignites could present some typical cellular plant structures such as fusinites and semifusinites, intergranular and plant tissue pores along with some crystalline impurities identified mainly as calcium carbonates and silicon dioxide as well as other trace elements such as aluminum and iron ions. This was confirmed by EDX analysis where the raw material presents high peak intensities of oxygen, silica and calcium (Figure 8a). After the adsorption process, SEM images of the metal-loaded lignite presented a clearer contrast on its surface for both metals, while the crystalline structure present in the raw material became less apparent (Figure 7b,c). This observation could be due to an adsorption reaction of metals on the lignite surface and at the same time to a possible inverse movement of some minerals from lignite to the liquid phase. The EDX analysis confirmed this hypothesis since the peaks related to oxygen, silica, potassium, and calcium decreased in intensity, while peaks attributed to copper and cadmium appeared, which confirms the adsorption process (Figure 8b,c).

The proximate analysis (Figure 9) indicates that the raw lignite is mainly composed of minerals with an ash content of about 39%. Its moisture, fixed carbon and volatile matter contents are respectively 7%, 21% and 33%. Similar trends were reported by Kanca [63] for a Turkish lignite, where ash content had the major proportion (48%) followed by volatile matter and fixed carbon (25% and 24%, respectively).

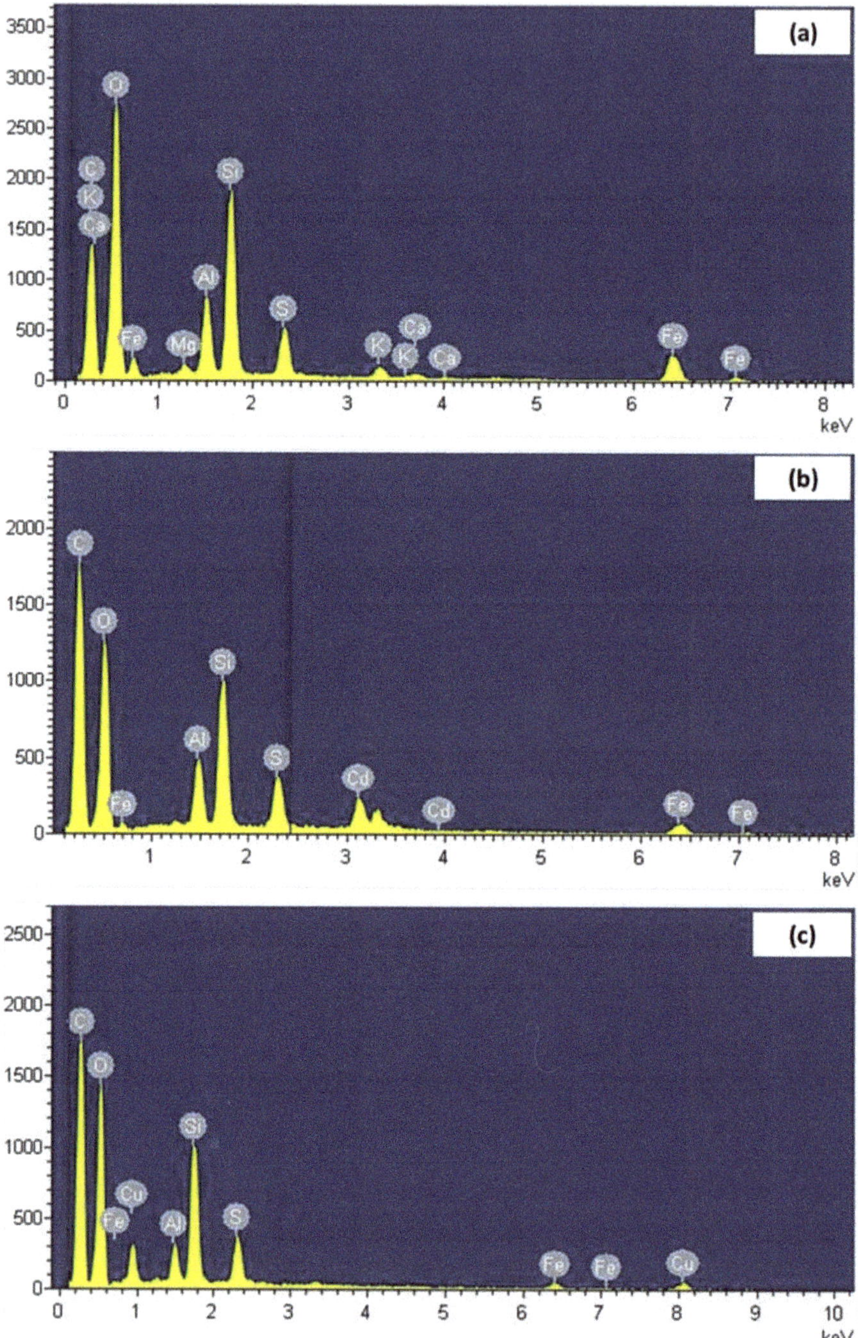

Figure 8. EDX analysis of: (**a**) raw lignite; (**b**) cadmium-loaded lignite; (**c**) copper-loaded lignite.

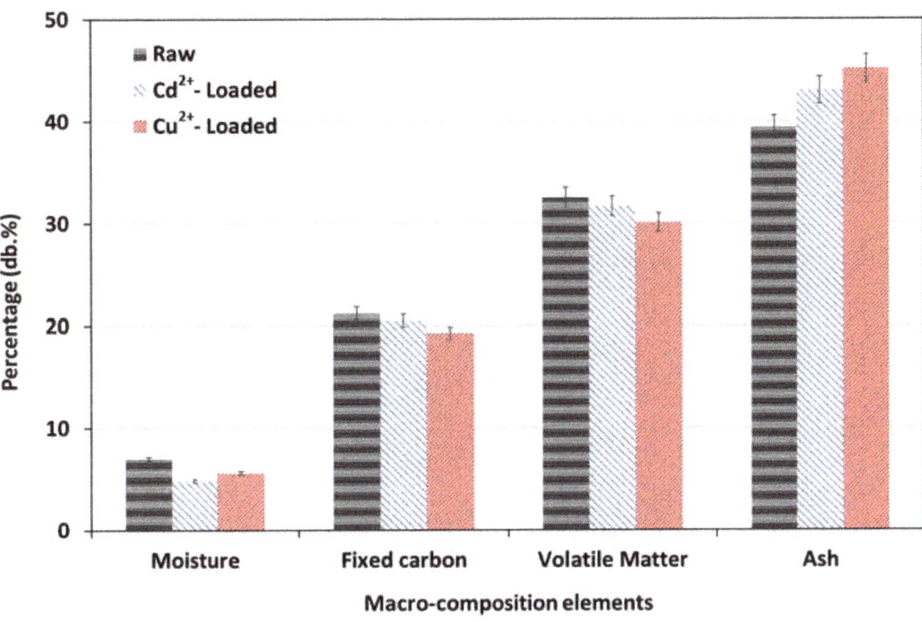

Figure 9. Proximate analysis of lignite before and after copper and cadmium adsorption.

After metal adsorption, a slight decrease of the fixed carbon and volatile matter contents occurred in favor of an ash content increase by about 3.7% and 5.8% for Cd- and Cu-loaded lignite, respectively. To align these results with SEM/EDS observations, it is possible that the adsorption of cadmium and copper were driven by three mechanisms: (i) adsorption of a small fraction of metal ions onto lignin decayed matrix which constitutes the small specific surface area of the feedstock yet characterized, (ii) ion-exchange reaction where ions such as potassium, calcium then silica at lower percentage were diffused from the solid matrix to the aqueous solution thus vacating the oxygenic functional groups, and (iii) chemical adsorption on free surface functional groups.

In order to confirm this hypothesis, FTIR analyses were performed on both raw and metal-loaded lignite and results are depicted in Figure 10 and Table 4. The lignite presented a heterogeneous surface with the co-presence of acidic and alkaline functional groups namely, hydroxyl (–OH; 3600–3200 cm^{-1}), aliphatic (C–H; 3000–2700 cm^{-1}), carbonyl and acetyl esters (C=O and C–O; 1650–1600 cm^{-1} and 1180–980 cm^{-1}, respectively), methyl and methylene aromatic (–CH$_2$/–CH$_3$; 1465–1320 cm^{-1}) and out-of-plane aromatic groups (C–H; 896–809 cm^{-1}). After metal adsorption, few modifications were noticed for peak positions of some functional groups due to their involvement in the adsorption process (Table 4). For instance, a peak shift of about +8 and +4 cm^{-1} was noticed for hydroxyl groups after Cd and Cu adsorption, respectively. Similar observation for carboxylic group where a vibration was spotted by about −4 and −6 cm^{-1} as compared to raw lignite for the same exhausted adsorbents, respectively. Moreover, a significant change was observed in the—C=C aromatic structure where a peak appeared for both Cd- and Cu-related specters at 1382 and 1380 cm^{-1}, respectively (Table 4). This could be related to the uptake of nitrate ions (entering in the composition of copper and cadmium reagents) by the lignite matrix, causing a slight alteration in its surface functionalities [64].

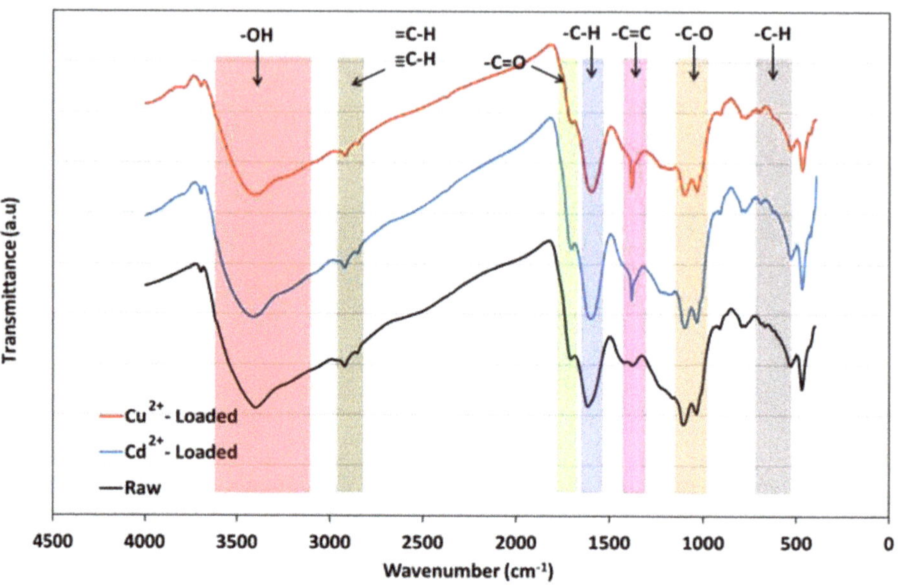

Figure 10. FTIR analysis of lignite before and after Cu (II) and Cd (II) adsorption.

Table 4. Peak location for each functional group detected in FTIR spectra for raw and metal-loaded lignite.

Functional Group	–OH	C=O	C–H	–C=C	–C–O	–C–H
Raw	3400	1705	1614	Absent	1105	794
Cd^{2+}—loaded lignite	3408	1701	1600	1382	1101	794
Cu^{2+}—loaded lignite	3404	1699	1593	1380	1103	794

On the other hand, the presence of positively charged metals (i.e., Cd^{2+} and Cu^{2+}) caused a thermodynamic balance between solid and liquid phases. It is possible that a cation release phenomenon occurred leading to the release of protons (H^+) or other cations (Na^+; K^+; Ca^{2+}, and Mg^{2+}) followed by the fixation of bivalent metals on the oxygenic functional groups as follows [19,65]:

$$2\left[A - C - O^- - H^+\right] + Me^{2+} \leftrightarrow \left[A - C - O^- - Me^{2+}\right] + 2H^+ \tag{5}$$

$$2\left[A - O^- - H^+\right] + Me^{2+} \leftrightarrow \left[A - O^- - Me^{2+}\right] + 2H^+ \tag{6}$$

$$\left[A - Na^+/K^+/Ca^{2+}/Mg^{2+}\right] + Me^{2+} \leftrightarrow \left[A - Me^{2+}\right] + 2Na^+/2K^+/Ca^{2+}/Mg^{2+} \tag{7}$$

where A stands for the aromatic structure of lignite and Me^{2+} is either Cu^{2+} or Cd^{2+}.

On the basis of FTIR investigation, we can deduce that Cd and Cu removal by lignite is governed not only by its textural properties (especially surface area and microporosity), but also by its richness in functional groups.

4. Conclusions

The current research study demonstrates that lignite, as a low cost and abundant material, can achieve very rapid and effective removal of cadmium and copper from aqueous effluents under wide experimental conditions. The removal efficiency seems not only dependent on the heavy metal properties but also on the textural and structural characteristics of the lignite. The heavy metal adsorption process occurs through a combination of several mechanisms, including mainly cation exchange and complexation with various functional groups. Lignite adsorption capacity could be further enhanced through physical, chemical, and thermal treatment methods. However, the cost and the environmental impact of such modifications should be accurately assessed. Moreover, further studies are required in order to assess the efficiency of raw/modified lignite for the removal of other heavy metals and recalcitrant organic pollutants under dynamic conditions.

Author Contributions: Conceptualization, S.J. and A.M.; methodology, S.J., A.M. and M.J.; validation, S.J., A.M. and A.A.A.; formal analysis, S.J., A.A.A., M.J., H.H. and A.M.; investigation, S.J. and A.M.; resources, S.J. and A.M.; writing—original draft preparation, S.J. and A.A.A.; writing—review and editing, A.M., H.H. and M.J.; visualization, S.J., A.M., A.A.A., H.H. and M.J.; supervision, S.J. and A.M.; project administration, A.M.; funding acquisition, A.M. All authors have read and agreed to the published version of the manuscript.

Funding: This research was funded by The Tunisian Ministry of Higher Education and Scientific Research.

Institutional Review Board Statement: Not applicable.

Informed Consent Statement: Not applicable.

Data Availability Statement: Data available on request.

Acknowledgments: Authors would like to thank all technicians involved in this work. The Tunisian Ministry of Higher Education and Scientific Research is gratefully acknowledged for financing this research project.

Conflicts of Interest: The authors declare no conflict of interest

References

1. Morais, S.; Costa, F.G.; Pereir, M.L. Heavy Metals and Human Health. In *Environmental Health—Emerging Issues and Practice*; IntechOpen: London, UK, 2012.
2. Sharma, H.; Rawal, N.; Mathew, B.B. The Characteristics, Toxicity and Effects of Cadmium. *Int. J. Nanotechnol. Nanosci.* **2015**, *3*, 1–9.
3. Araya, M.; Olivares, M.; Pizarro, F. Copper in human health. *Int. J. Environ. Health* **2007**, *1*, 608. [CrossRef]
4. Abdullah, N.; Yusof, N.; Lau, W.; Jaafar, J.; Ismail, A. Recent trends of heavy metal removal from water/wastewater by membrane technologies. *J. Ind. Eng. Chem.* **2019**, *76*, 17–38. [CrossRef]
5. Liu, Y.; Ke, X.; Zhu, H.; Chen, R.; Chen, X.; Zheng, X.; Jin, Y.; Van Der Bruggen, B.; Jin, Y. Treatment of raffinate generated via copper ore hydrometallurgical processing using a bipolar membrane electrodialysis system. *Chem. Eng. J.* **2020**, *382*, 122956. [CrossRef]
6. Sun, Y.; Zhou, S.; Pan, S.-Y.; Zhu, S.; Yu, Y.; Zheng, H. Performance evaluation and optimization of flocculation process for removing heavy metal. *Chem. Eng. J.* **2020**, *385*, 123911. [CrossRef]
7. Cui, Y.; Ge, Q.; Liu, X.-Y.; Chung, N.T.-S. Novel forward osmosis process to effectively remove heavy metal ions. *J. Membr. Sci.* **2014**, *467*, 188–194. [CrossRef]
8. Duan, C.; Ma, T.; Wang, J.; Zhou, Y. Removal of heavy metals from aqueous solution using carbon-based adsorbents: A review. *J. Water Process. Eng.* **2020**, *37*, 101339. [CrossRef]
9. Qin, H.; Hu, T.; Zhai, Y.; Lu, N.; Aliyeva, J. The improved methods of heavy metals removal by biosorbents: A review. *Environ. Pollut.* **2020**, *258*, 113777. [CrossRef]
10. Shakoor, M.B.; Ali, S.; Rizwan, M.; Abbas, F.; Bibi, I.; Riaz, M.; Khalil, U.; Niazi, N.K.; Rinklebe, J. A review of biochar-based sorbents for separation of heavy metals from water. *Int. J. Phytoremediation* **2020**, *22*, 111–126. [CrossRef]
11. Baby, R.; Saifullah, B.; Rehman, F.U.; Shaikh, R.I. Greener Method for the Removal of Toxic Metal Ions from the Wastewater by Application of Agricultural Waste as an Adsorbent. *Water* **2018**, *10*, 1316. [CrossRef]
12. Cataldo, S.; Gianguzza, A.; Pettignano, A.; Piazzese, D.; Sammartano, S. Complex formation of copper(II) and cadmium(II) with pectin and polygalacturonic acid in aqueous solution. An ISE-H + and ISE-Me 2+ electrochemical study. *Int. J. Electrochem. Sci.* **2012**, *7*, 6722–6737.
13. Zhong-Kai, Y.; Chen, Y.-D.; Yang, Z.-K.; Nagarajan, D.; Chang, J.-S.; Ren, N.-Q. High-efficiency removal of lead from wastewater by biochar derived from anaerobic digestion sludge. *Bioresour. Technol.* **2017**, *246*, 142–149. [CrossRef]
14. Chouchene, A.; Jeguirim, M.; Trouvé, G. Biosorption performance, combustion behavior, and leaching characteristics of olive solid waste during the removal of copper and nickel from aqueous solutions. *Clean Technol. Environ. Policy* **2013**, *16*, 979–986. [CrossRef]

15. Mlayah, A.; Jellali, S. Study of continuous lead removal from aqueous solutions by marble wastes: Efficiencies and mechanisms. *Int. J. Environ. Sci. Technol.* **2014**, *12*, 2965–2978. [CrossRef]
16. Li, G.; Zhang, J.; Liu, J.; Sun, C.; Yan, Z. Adsorption characteristics of white pottery clay towards Pb(II), Cu(II), and Cd(II). *Arab. J. Geosci.* **2020**, *13*, 519. [CrossRef]
17. Foroutan, R.; Mohammadi, R.; Farjadfard, S.; Esmaeili, H.; Saberi, M.; Sahebi, S.; Dobaradaran, S.; Ramavandi, B. Characteristics and performance of Cd, Ni, and Pb bio-adsorption using Callinectes sapidus biomass: Real wastewater treatment. *Environ. Sci. Pollut. Res.* **2019**, *26*, 6336–6347. [CrossRef]
18. Jellali, S.; Diamantopoulos, E.; Haddad, K.; Anane, M.; Durner, W.; Mlayah, A. Lead removal from aqueous solutions by raw sawdust and magnesium pretreated biochar: Experimental investigations and numerical modelling. *J. Environ. Manag.* **2016**, *180*, 439–449. [CrossRef]
19. Puglla, E.P.; Guaya, D.; Tituana, C.; Osorio, F.; García-Ruiz, M.J. Biochar from Agricultural By-Products for the Removal of Lead and Cadmium from Drinking Water. *Water* **2020**, *12*, 2933. [CrossRef]
20. Egirani, D.E.; Poyi, N.R.; Shehata, N. Preparation and characterization of powdered and granular activated carbon from Palmae biomass for cadmium removal. *Int. J. Environ. Sci. Technol.* **2020**, *17*, 2443–2454. [CrossRef]
21. Vo, A.T.; Nguyen, V.P.; Ouakouak, A.; Nieva, A.; Doma, J.B.T.; Tran, H.N.; Chao, H.-P. Efficient Removal of Cr(VI) from Water by Biochar and Activated Carbon Prepared through Hydrothermal Carbonization and Pyrolysis: Adsorption-Coupled Reduction Mechanism. *Water* **2019**, *11*, 1164. [CrossRef]
22. Ediger, V.Ş.; Berk, I.; Ersoy, M. An assessment of mining efficiency in Turkish lignite industry. *Resour. Policy* **2015**, *45*, 44–51. [CrossRef]
23. Kuśmierek, K.; Sprynskyy, M.; Świątkowski, A. Raw lignite as an effective low-cost adsorbent to remove phenol and chlorophenols from aqueous solutions. *Sep. Sci. Technol.* **2020**, *55*, 1741–1751. [CrossRef]
24. Grajales-Mesa, S.J.; Malina, G. Pilot-Scale Evaluation of a Permeable Reactive Barrier with Compost and Brown Coal to Treat Groundwater Contaminated with Trichloroethylene. *Water* **2019**, *11*, 1922. [CrossRef]
25. Zhang, W.; Gago-Ferrero, P.; Gao, Q.; Ahrens, L.; Blum, K.; Rostvall, A.; Björlenius, B.; Andersson, P.L.; Wiberg, K.; Haglund, P.; et al. Evaluation of five filter media in column experiment on the removal of selected organic micropollutants and phosphorus from household wastewater. *J. Environ. Manag.* **2019**, *246*, 920–928. [CrossRef]
26. Nazari, M.A.; Mohaddes, F.; Pramanik, B.K.; Othman, M.; Muster, T.; Bhuiyan, M.A. Application of Victorian brown coal for removal of ammonium and organics from wastewater. *Environ. Technol.* **2018**, *39*, 1041–1051. [CrossRef]
27. Zherebtsov, S.I.; Malyshenko, N.V.; Votolin, K.S.; Ismagilov, Z.R. Sorption of Metal Cations by Lignite and Humic Acids. *Coke Chem.* **2020**, *63*, 142–148. [CrossRef]
28. Binabaj, M.A.; Nowee, S.M.; Ramezanian, N. Comparative study on adsorption of chromium(VI) from industrial wastewater onto nature-derived adsorbents (brown coal and zeolite). *Int. J. Environ. Sci. Technol.* **2017**, *15*, 1509–1520. [CrossRef]
29. Zhao, T.-T.; Ge, W.-Z.; Yue, F.; Wang, Y.; Pedersen, C.M.; Zeng, F.-G.; Qiao, Y. Mechanism study of Cr(III) immobilization in the process of Cr(VI) removal by Huolinhe lignite. *Fuel Process. Technol.* **2016**, *152*, 375–380. [CrossRef]
30. Pentari, D.; Perdikatsis, V.; Katsimicha, D.; Kanaki, A. Sorption properties of low calorific value Greek lignites: Removal of lead, cadmium, zinc and copper ions from aqueous solutions. *J. Hazard. Mater.* **2009**, *168*, 1017–1021. [CrossRef]
31. Uçurum, M. A study of removal of Pb heavy metal ions from aqueous solution using lignite and a new cheap adsorbent (lignite washing plant tailings). *Fuel* **2009**, *88*, 1460–1465. [CrossRef]
32. Pehlivan, E.; Arslan, G. Removal of metal ions using lignite in aqueous solution—Low cost biosorbents. *Fuel Process. Technol.* **2007**, *88*, 99–106. [CrossRef]
33. Doskočil, L.; Pekař, M. Removal of metal ions from multi-component mixture using natural lignite. *Fuel Process. Technol.* **2012**, *101*, 29–34. [CrossRef]
34. Pehlivan, E.; Richardson, A.; Zuman, P. Electrochemical Investigation of Binding of Heavy Metal Ions to Turkish Lignites. *Electroanalysis* **2004**, *16*, 1292–1298. [CrossRef]
35. Azzaz, A.A.; Jellali, S.; Assadi, A.A.; Bousselmi, L. Chemical treatment of orange tree sawdust for a cationic dye enhancement removal from aqueous solutions: Kinetic, equilibrium and thermodynamic studies. *Desalination Water Treat.* **2015**, *57*, 22107–22119. [CrossRef]
36. Azzaz, A.A.; Jeguirim, M.; Kinigopoulou, V.; Doulgeris, C.; Goddard, M.-L.; Jellali, S.; Ghimbeu, C.M. Olive mill wastewater: From a pollutant to green fuels, agricultural and water source and bio-fertilizer—Hydrothermal carbonization. *Sci. Total. Environ.* **2020**, *733*, 139314. [CrossRef] [PubMed]
37. Haddad, K.; Jeguirim, M.; Jerbi, B.; Chouchene, A.; Dutournié, P.; Thevenin, N.; Ruidavets, L.; Jellali, S.; Limousy, L. Olive Mill Wastewater: From a Pollutant to Green Fuels, Agricultural Water Source and Biofertilizer. *ACS Sustain. Chem. Eng.* **2017**, *5*, 8988–8996. [CrossRef]
38. Havelcová, M.; Mizera, J.; Sýkorová, I.; Pekař, M. Sorption of metal ions on lignite and the derived humic substances. *J. Hazard. Mater.* **2009**, *161*, 559–564. [CrossRef]
39. Huber, M.; Badenberg, S.C.; Wulff, M.; Drewes, J.E.; Helmreich, B. Evaluation of Factors Influencing Lab-Scale Studies to Determine Heavy Metal Removal by Six Sorbents for Stormwater Treatment. *Water* **2016**, *8*, 62. [CrossRef]
40. Kumar, P.S.; Korving, L.; Keesman, K.J.; Van Loosdrecht, M.C.; Witkamp, G.-J. Effect of pore size distribution and particle size of porous metal oxides on phosphate adsorption capacity and kinetics. *Chem. Eng. J.* **2019**, *358*, 160–169. [CrossRef]

41. Pn, I.; Cp, U. Overview on the Effect of Particle Size on the Performance of Wood Based Adsorbent. *J. Biosens. Bioelectron.* **2016**, *7*, 1–4. [CrossRef]
42. Jellali, S.; Wahab, M.A.; Anane, M.; Riahi, K.; Jedidi, N. Biosorption characteristics of ammonium from aqueous solutions onto Posidonia oceanica (L.) fibers. *Desalination* **2011**, *270*, 40–49. [CrossRef]
43. Pentari, D.; Vamvouka, D. Cadmium Removal from Aqueous Solutions Using Nano-Iron Doped Lignite. *IOP Conf. Ser. Earth Environ. Sci.* **2019**, *221*, 012135. [CrossRef]
44. Shrestha, S.; Son, G.; Lee, S.H. Kinetic Parameters and Mechanism of Zn (II) Adsorption on Lignite and Coconut Shell–Based Activated Carbon Fiber. *J. Hazard. Toxic Radioact. Waste* **2014**, *18*, 1–8. [CrossRef]
45. Pardo-Botello, R.; Fernández-González, C.; Pinilla-Gil, E.; Cuerda-Correa, E.M.; Gómez-Serrano, V. Adsorption kinetics of zinc in multicomponent ionic systems. *J. Colloid Interface Sci.* **2004**, *277*, 292–298. [CrossRef]
46. Jellali, S.; Wahab, M.A.; Ben Hassine, R.; Hamzaoui, A.H.; Bousselmi, L. Adsorption characteristics of phosphorus from aqueous solutions onto phosphate mine wastes. *Chem. Eng. J.* **2011**, *169*, 157–165. [CrossRef]
47. Haddad, K.; Jellali, S.; Jaouadi, S.; Benltifa, M.; Mlayah, A.; Hamzaoui, A.H. Raw and treated marble wastes reuse as low cost materials for phosphorus removal from aqueous solutions: Efficiencies and mechanisms. *Comptes Rendus Chim.* **2015**, *18*, 75–87. [CrossRef]
48. Haddad, K.; Jellali, S.; Jeguirim, M.; Trabelsi, A.B.H.; Limousy, L. Investigations on phosphorus recovery from aqueous solutions by biochars derived from magnesium-pretreated cypress sawdust. *J. Environ. Manag.* **2018**, *216*, 305–314. [CrossRef]
49. Gürses, A.; Hassani, A.; Kıranşan, M.; Açışlı, Ö.; Karaca, S. Removal of methylene blue from aqueous solution using by untreated lignite as potential low-cost adsorbent: Kinetic, thermodynamic and equilibrium approach. *J. Water Process. Eng.* **2014**, *2*, 10–21. [CrossRef]
50. Allen, S.J.; Brown, P.A. Isotherm analyses for single component and multi-component metal sorption onto lignite. *J. Chem. Technol. Biotechnol.* **1995**, *62*, 17–24. [CrossRef]
51. Karabulut, S.; Karabakan, A.; Denizli, A.; Yürüm, Y. Batch removal of copper(II) and zinc(II) from aqueous solutions with low-rank Turkish coals. *Sep. Purif. Technol.* **2000**, *18*, 177–184. [CrossRef]
52. Sun, H.; He, X.; Wang, Y.; Cannon, F.S.; Wen, H.; Li, X. Nitric acid-anionic surfactant modified activated carbon to enhance cadmium(II) removal from wastewater: Preparation conditions and physicochemical properties. *Water Sci. Technol.* **2018**, *78*, 1489–1498. [CrossRef] [PubMed]
53. Pekař, M.; Klučáková, M. Comparison of Copper Sorption on Lignite and on Soils of Different Types and Their Humic Acids. *Environ. Eng. Sci.* **2008**, *25*, 1123–1128. [CrossRef]
54. Singh, S.; Kumar, V.; Datta, S.; Dhanjal, D.S.; Sharma, K.; Samuel, J.; Singh, J. Current advancement and future prospect of biosorbents for bioremediation. *Sci. Total. Environ.* **2020**, *709*, 135895. [CrossRef] [PubMed]
55. Chakraborty, R.; Asthana, A.; Singh, A.K.; Jain, B.; Susan, A.B.H. Adsorption of heavy metal ions by various low-cost adsorbents: A review. *Int. J. Environ. Anal. Chem.* **2020**, 1–38. [CrossRef]
56. Leyva-Ramos, R.; Bernal-Jacome, L.; Acosta-Rodriguez, I. Adsorption of cadmium(II) from aqueous solution on natural and oxidized corncob. *Sep. Purif. Technol.* **2005**, *45*, 41–49. [CrossRef]
57. Krishnani, K.K.; Meng, X.; Christodoulatos, C.; Boddu, V.M. Biosorption mechanism of nine different heavy metals onto biomatrix from rice husk. *J. Hazard. Mater.* **2008**, *153*, 1222–1234. [CrossRef] [PubMed]
58. Demessie, B.; Sahle-Demessie, E.; Sorial, G.A. Cleaning Water Contaminated With Heavy Metal Ions Using Pyrolyzed Biochar Adsorbents. *Sep. Sci. Technol.* **2015**, *50*, 2448–2457. [CrossRef]
59. Villaescusa, I.; Fiol, N.; Martínez, M.; Miralles, N.; Poch, J.; Serarols, J. Removal of copper and nickel ions from aqueous solutions by grape stalks wastes. *Water Res.* **2004**, *38*, 992–1002. [CrossRef]
60. Zhang, C.; Li, J.; Chen, Z.; Cheng, F. Factors controlling adsorption of recalcitrant organic contaminant from bio-treated coking wastewater using lignite activated coke and coal tar-derived activated carbon. *J. Chem. Technol. Biotechnol.* **2017**, *93*, 112–120. [CrossRef]
61. Zhang, N.; Shen, Y. One-step pyrolysis of lignin and polyvinyl chloride for synthesis of porous carbon and its application for toluene sorption. *Bioresour. Technol.* **2019**, *284*, 325–332. [CrossRef]
62. Zhang, B.; Chen, Y. Particle size effect on pore structure characteristics of lignite determined via low-temperature nitrogen adsorption. *J. Nat. Gas Sci. Eng.* **2020**, *84*, 103633. [CrossRef]
63. Kanca, A. Investigation on pyrolysis and combustion characteristics of low quality lignite, cotton waste, and their blends by TGA-FTIR. *Fuel* **2020**, *263*, 116517. [CrossRef]
64. Qi, Y.; Hoadley, A.F.; Chaffee, A.L.; Garnier, G. Characterisation of lignite as an industrial adsorbent. *Fuel* **2011**, *90*, 1567–1574. [CrossRef]
65. Azzaz, A.A.; Jellali, S.; Bengharez, Z.; Bousselmi, L.; Akrout, H. Investigations on a dye desorption from modified biomass by using a low-cost eluent: Hysteresis and mechanisms exploration. *Int. J. Environ. Sci. Technol.* **2019**, *16*, 7393–7408. [CrossRef]

Article

Advanced Treatment of Real Grey Water by SBR Followed by Ultrafiltration—Performance and Fouling Behavior

Gabriela Kamińska and Anna Marszałek *

Department of Water and Wastewater Engineering, Silesian University of Technology, Konarskiego 18, 44-100 Gliwice, Poland; gabriela.liszczyk@gmail.com
* Correspondence: anna.marszalek@polsl.pl; Tel.: +48-32-237-21-73

Received: 22 November 2019; Accepted: 3 January 2020; Published: 4 January 2020

Abstract: Grey water has been identified as a potential source of water in a number of applications e.g., toilet flushing, laundering in first rinsing, floor cleaning, and irrigation. The major obstacle to the reuse of grey water relates to pathogens, nutrients, and organic matter found in grey water. Therefore, much effort has been put to treat grey water, in order to yield high-quality water deprived of bacteria and with an appropriate value in a wide range of quality parameters (Total Organic Carbon (TOC), nitrate, phosphate, ammonium, pH, and absorbance), similar to the values for tap water. The aim of this study was to treat the real grey water, and turn it into high-quality, safe water. For this purpose, the real grey water was treated by means of a sequential biological reactor (SBR) followed by ultrafiltration. Initially, grey water was treated in a laboratory SBR reactor with a capacity of 3 L, operated in a 24 h cycle. Then, SBR effluent was purified in a cross-flow ultrafiltration setup. Treatment efficiency in SBR and ultrafiltration was assessed using extended physicochemical and microbiological analyses (pH, conductivity, color, absorbance, Chemical Oxygen Demand (COD), Biological Oxygen Demand (BOD$_5$), nitrate, phosphate, ammonium, total nitrogen, phenol index, nonionic and anionic surfactants, TOC, *Escherichia coli*, and enterococci). Additionally, ultrafiltration was evaluated in terms of fouling behavior for three polymer membranes with different MWCO (molecular weight cut-off). The values of quality parameters (pH, conductivity, COD, BOD$_5$, TOC, N-NH$_4^+$, N-NO$_3^-$, N$_{tot}$, and P-PO$_4^{3-}$) measured in SBR effluent did not exceed permissible values for wastewater discharged to soil and water. Ultrafiltration provided the high-quality water with very low values of COD (5.8–18.1 mg/L), TOC (0.47–2.19 mg/L), absorbance$_{UV254}$ (0.015–0.048 1/cm), color (10–29 mgPt/L) and concentration of nitrate (0.18–0.56 mg/L), phosphate (0.9–2.1 mg/L), ammonium (0.03–0.11 mg/L), and total nitrogen (3.3–4.7 mg/L) as well as lack of *E. coli* and enterococci. Membrane structural and surface properties did not affect the treatment efficiency, but did influence the fouling behavior.

Keywords: grey water; SBR; ultrafiltration; fouling; zeta potential

---

1. Introduction

Water is emerging as one of the single most important resources of Planet Earth for the prosperity of the economy and human life. However, freshwater resources have been increasingly polluted and depleted globally. The constant increase in water usage and climate change—such as altered weather-patterns (including droughts or floods), deforestation, and increased pollution—are the main driving forces for the rising global scarcity of water. It is experienced especially by European countries, due to an improvement of living standards and economic development in recent years. Along with economic development, the total water use in Europe has been sharply increasing in the last decades [1]. Grey water recycling is the solution to the growing challenges associated with water shortages [2]. Quantitatively, grey water represents around 65% of the total volume of domestic sewage, making it the largest stream for water reuse [3]. Grey water contains nitrate, phosphate, organic matter, surfactants, pharmaceuticals, oils, and pathogens [4]. Therefore, the treatment of grey water before its reuse is necessary. Purified grey water could be reused for cleaning, car washing, concrete production, and irrigation [5,6]. The guidelines of the World Health Organization (WHO) point to four criteria for the reuse of grey water: hygiene, aesthetics, environmental tolerance, and economic feasibility. WHO recommends biological treatment and ultrafiltration in order to obtain high-quality water. However, the requirements that define "high water quality" are too sweeping and do not show the physical and chemical specifications. It is caused by a number of possible applications of reclaimed grey water that require various different quality standards. In 2006, WHO set microbiological criteria that reclaimed grey water should meet for its reuse for restricted and non-restricted agricultural irrigation. [7,8].

Numerous approaches, including low and high pressure-driven membrane techniques [9–11], coagulation [12,13], biological processes, and membrane bioreactors [3], have been proposed for the treatment of grey water. These technologies vary in both complexity and performance. For example, Li et al. recommended the employment of an aerobic treatment coupled with membrane filtration or an aerobic treatment with sand filtration, with disinfection as the last step for both systems [8]. Similarly, Ding et al. suggested applying a system that combines membrane filtration with biological treatment in a gravity-driven membrane filtration system. In this technique, the fouling layer attached to the membrane stabilizes a flux and improves the treatment effects due to the microbial activity of biofilm [14]. From these studies, it is clear that grey water requires both biological and physical treatment to meet non-potable reuse standards. Aerobic biological processes are effective to remove organics from grey water. Physical processes, particularly membrane filtration, are recommended for polishing effluents from the biological treatment step [9,15]. Ultrafiltration (UF) retains bacteria, suspension, colloids, natural organic matter, and partial micropollutants such as pharmaceuticals and personal care products. However, the efficiency of ultrafiltration in the removal of individual pollutants depends on membrane type and its properties such as molecular weight cut-off (MWCO). Membrane properties such as contact angle and zeta potential play an important role in the fouling behavior and hydraulic performance of the system [16]. Membrane fouling results in a permeability loss due to the increase in hydraulic resistances in the filtration system, especially in the case of porous polymer membranes. There are many studies reported in the literature that aimed at mitigating the fouling by selecting the most optimal operational parameters for ultrafiltration such as transmembrane pressure, velocity, temperature, or developing a cleaning method. [17–19]. On the other hand, they do not consider the effect of structure and surface properties of membrane i.e., MWCO and contact angle on fouling behavior. Another issue is that most of these studies were conducted for artificial grey water. While artificial grey water provides useful information for model development, it does not reflect real conditions, and the composition affects the process significantly.

The main novelty of this work shows that high-quality water can be obtained from grey water. For a need of this study, we established that high-quality water is water: (1) that fulfills criteria included in the Polish Regulation from the Minister of the Environment on the conditions to be met when discharging wastewater into water or soil and (2) with the basic quality parameters (smell, color, turbidity, TOC, ammonium, nitrate, chloride, conductivity, hardness, *E. coli*, and enterococci) like the standards for tap water. In order to gain this purpose, real grey water was treated using a sequential biological reactor (SBR) followed by ultrafiltration. Treatment efficiency was evaluated for SBR and ultrafiltration using extended physicochemical and microbiological analyses. Additionally, ultrafiltration was studied in terms of fouling behavior for three polymer membranes with different MWCO. In that context, the novelty of this work is also a determination of the effect of membrane type on the treatment of SBR effluent and fouling behavior.

2. Methodology

2.1. Grey Water Characteristic

Grey water was collected from a single-family household located in Silesia in Poland. This house is equipped with an installation to collect grey water from the shower, kitchen sink, and washing machine. The characteristic of the grey water is given in Table 1. Samples were taken directly from a storage tank and transported to the lab facilities, stored at 4 °C and analyzed within 48 h according to the methodology presented in Section 2.6.

Table 1. Characteristic of grey water, as taken.

Parameter	Unit	Value		
		Min	Max	Average
pH	-	7.47	7.88	7.6
Conductivity	μS/cm	735	772	758
Color	mgPt/L	182	228	201
Turbidity	FTU	26	31	28.2
Absorbance$_{UV254}$	1/cm	0.468	0.611	0.55
COD	mg/L	415	638	543
BOD$_5$	mg/L	190	240	212
N-NO$_3^-$	mg/L	0.4	0.8	0.6
P-PO$_4^{3-}$	mg/L	3.8	4.6	4.15
N-NH$_4^+$	mg/L	0	0.3	0.2
N$_{tot}$	mg/L	9.4	11.3	10.7
Surfactants nonionic	mg/L	1.33	1.64	1.42
Surfactants anionic	mg/L	12.9	17.5	15.2
TOC	mg/L	51.2	59.3	56.3

2.2. Biological Treatment in SBR

The SBR technology used provides for the dosing of wastewater in alternately used anaerobic reactor oxygen conditions. This guarantees a high degree of removal of carbon, nitrogen, and phosphorus compounds. The operation of the reactor is simple, and during operation, modifications of individual separate phases of the technological cycles are possible. The biological process was carried out under laboratory conditions, using activated sludge taken from the municipal sewage treatment plant in Gliwice, Poland. The treatment of grey water was carried out in a laboratory SBR with a capacity of 3 L. During the treatment experiment, the excessive activated sludge was periodically removed from the SBR in order to keep its concentration at the level of 3.0 g/L. The solid retention time (SRT) was 20 days. The load of the sludge with the contaminants was equal to 0.1 g COD/g$_{DM}$d and the concentration of oxygen was at the level of 3 mg/L. The system was operated as the sequential biological reactor in one cycle per day. The length of particular operation stages was as follows: filling and mixing phase for 2 h, aeration phase for 21 h, and sedimentation and SBR effluent removal for 1 h. The aeration system consisted of aquarium cubes placed on the bottom of the tank with connected aeration pumps of the Tetratec APS 300 type (Tetra, Melle, Germany). In order to thoroughly mix the contents of the chamber, it was mixed by means of a magnetic stirrer with the possibility of speed regulation in the range from 50 to 1000 revolutions/min. The chamber was filled and emptied using Heidolph peristaltic pumps (Heidolph, Schwabach, Germany). The electronic weekly programmer controlled the start and end of the relevant SBR work phase.

2.3. SBR Effluent Treatemnt by Ultrafilltration–Ultrafiltration Run

Ultrafiltration is especially dedicated to drinking water production because colloids, particulates, macromolecules, and pathogens are removed in this process. It is also a suitable process to purify effluent after biological treatment as this effluent contains bacteria, particles of organic matter, and suspended solids. Ultrafiltration was carried out in the lab, using a scale cross-flow configuration equipped with a plate-and-frame membrane module SEPA CF-NP (GE Osmonics, Minnetonka, MN, USA) as seen in Figure 1. Three ultrafiltration membranes were used in separate processes. Membrane properties are presented in Table 2. Before each experiment, the clean water flux was determined using ultrapure water. The process was operated at a constant pressure of 5 bar and a constant temperature of 22 ± 1 °C and a constant velocity of 1 m/s with continuous dosing of SBR effluent to the feed tank. Each filtration run consisted of three cycles including 60 min of filtration followed by forward flushing with ultrapure water during 60 s. The volume of permeate was monitored in order to determine the permeability from the following equation:

$$L_p = \frac{V}{\Delta p \cdot A \cdot t}$$

(1)

where: L_p is permeability ($L \cdot m^{-2} \cdot h^{-1} \cdot bar^{-1}$) in short (LMHB), V is permeate volume (L), A is membrane surface area (m^2), t is permeate time collection (h), and Δp is transmembrane pressure (bar).

Table 2. Properties of ultrafiltration membranes.

Symbol	DSGM	V3	BN
Manufacturer	GE Osmonics (Minnetonka, MN, USA)	Synder (Allison Parkway Vacaville, CA USA)	Synder
Polymer material	Polyamide-TFC	PVDF	PVDF
MWCO, Da	8000	30,000	50,000
Thickness *, µm	150	220	200

* own measurements with an electronic micrometer.

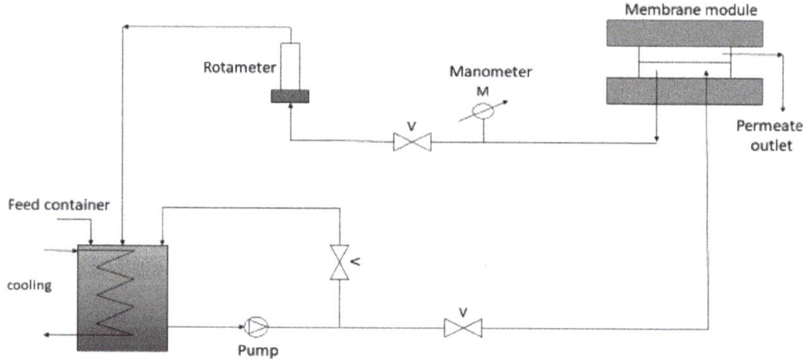

Figure 1. Schematic diagram of the cross-flow filtration set-up.

2.4. Membrane Fouling Characterization

In this study, hydraulic resistances were used to characterize the fouling behavior of ultrafiltration membranes treating SBR effluent. Hydraulic resistances of membrane and fouling layer were calculated using the resistance in the series model and Darcy's law using the correlations as shown below [20].

$$k_{tot} = k_m + k_f$$

(2)

$$k_f = k_{irr} + k_{rev}$$

(3)

$$J = \frac{\Delta p}{\mu \cdot k}$$

(4)

where: k is hydraulic resistance, where subscripts m, f, irr, rev, and tot are related to membrane, fouling, hydraulically irreversible fouling, hydraulically reversible fouling, and total (m^{-1}), respectively, J is the flux ($m^3 \cdot m^{-2} \cdot s^{-1}$), Δp is the transmembrane pressure ($kg \cdot s^{-2} \cdot m^{-1}$), and μ is the dynamic viscosity of water at given temperature ($kg \cdot m^{-1} \cdot s^{-1}$). Membrane resistance ($k_m$) was measured for the clean membrane with ultrapure water prior to feed water filtration. Hydraulically irreversible fouling was determined from the flux after forward flushing, while hydraulically reversible fouling was determined from the difference in fouling and irreversible resistances.

2.5. Membrane Characterization

Zeta potential of membranes was determined by the electrokinetic analyzer SurPASS ™ 3 (Anton Paar, Graz, Austria). Measurements of the contact angle were performed using the goniometer PG-1 (Fibro System AB, Hägersten, Sweden) and the sessile drop method was applied. By syringe on top, a drop of ultrapure water was put on the dried membrane surface. Through an enlarged projection of the water drop on the gauge, the value of the contact angle was measured. For every type of membrane, 10 samples were measured and the average value was calculated.

2.6. Quality Analysis and Microbiological Assessment

The treatment efficiency in SBR and ultrafiltration was evaluated by the monitoring of the typical quality parameters (color, turbidity, COD, BOD_5, TOC, phenolic index, absorbance of UV_{254}, nitrate ($N-NO_3^-$), phosphate ($P-PO_4^{3-}$), ammonium ($N-NH_4^+$), total nitrogen (N_{tot}), conductivity, pH, and anionic and non-ionic surfactants). Color and turbidity measurements were performed with a UV-Vis Spectroquant®Pharo 300 (Merck, Kenilworth, NJ, USA). Phenolic index, COD, nitrate, phosphate, ammonium, total nitrogen, and anionic and non-ionic surfactant concentrations were determined spectrophotometrically with Merck test kits (codes of the Merck test kits are given in Table S1). The absorbance was measured at 254 nm, using a UV-visible light (UV-Vis) Cecil 1000 (Analytik Jena AG company, Jena, Germany). TOC was measured using a TOC-L series analyzer (Shimadzu, Kioto, Prefektura Kioto, Japan). pH and conductivity were monitored by multifunctional analyzer CX-461 (Elmetron, Zabrze, Poland). The BOD_5 was determined by respirometric measurement using the OXI Top System WTW set (Xylem Analytics, Weilheim Germany). Microbiological analysis including *E. coli* and enterococci was conducted by an external accredited lab according to ISO methods PN-EN ISO 9308-1:2014-12/PN-EN ISO 9308-1/A1:2017; PN-EN ISO 7899-2:2004.

3. Results and Discussion

3.1. Grey Water Treatment in SBR

3.1.1. Reduction of COD, BOD_5, and TOC

The biological sewage treatment aimed to reduce biodegradable organic matter and remove the biogenic substances i.e., nitrogen and phosphorus. In the preliminary stage, the activated sludge was taken from the municipal wastewater plant, then it was adapted to a new influent (grey water). The results discussed in Figures 2–4 show the analysis of individual parameters after the adaptation process. Figure 2 shows the SBR system performance in relation to the concentration of organic compounds in purified grey water. Detailed results of physicochemical analyses are presented in Table S2 (in Supplementary file).

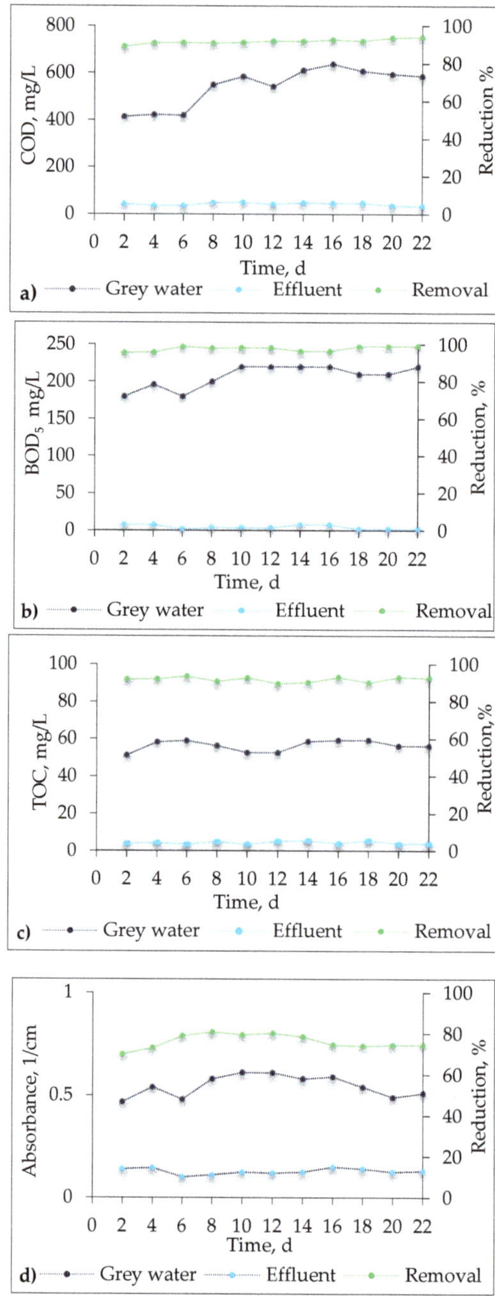

Figure 2. The performance of the sequential biological reactor (SBR) system with respect to the changes in the concentration of the organic compounds.

It was shown that the degree of pollution removal was high, and the values of the quality parameters in SBR effluent did not exceed the permissible values included in the currently in force in Poland, Regulation of the Minister of the Environment on the conditions to be met when discharging wastewater into water or soil [21]. Over the whole experimental period (22 days), the SBR system reduced COD by around 92%, from an average value of 543 mg/L (influent) to 44 mg/L (effluent) (Figure 2a). Fountoulakis et al. obtained a similar result in their studies [22]. More specifically, the removal of COD was approximately 87%. Such, a high removal was also obtained in studies on grey water purification in a membrane bioreactor operating in a flow-through system [3], where the COD reduction degree was 88%.

It was found that the easily biodegradable compounds expressed by BOD_5 were removed in 99% of cases, and the average concentration in the effluent was 4 mg/L (Figure 2b). Similarly, a high reduction was recorded for TOC and absorbance, i.e., 91% and 76%, respectively (Figure 2c,d). Anionic and nonionic surfactants were reduced by 97% and 100%, respectively (as seen in Table S1 in the Supplementary File). In the study [22], the removal efficiency of the anionic surfactants was about 80% in the submerged membrane bioreactor. This is due to the fact that non-ionic and anionic surfactants are readily biodegradable under aerobic conditions [23–25].

3.1.2. Removal of Biogenic Compounds

As seen in Figure 3, the concentration of nutrients in SBR effluent was low and did not exceed permissible values laid out in the Polish regulations [21].

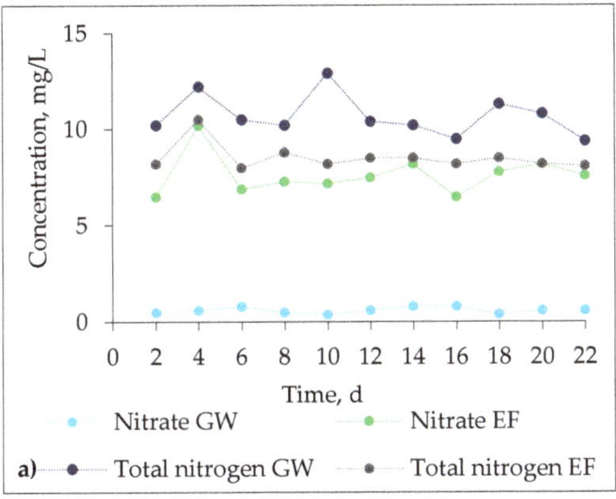

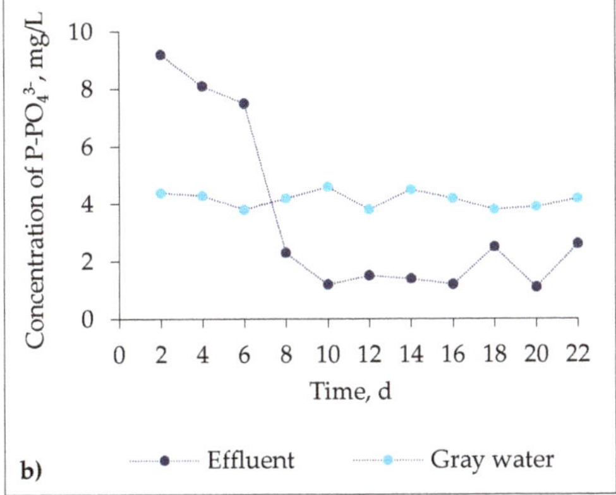

Figure 3. The performance of the SBR system with respect to the changes in the concentration of the nutrients. GW—grey water, EF—effluent.

Raw grey water was characterized by a minimum content of ammonium and nitrate nitrogen at the level of 0.2 mg/L and 0.8 mg/L, respectively. After biological treatment with the SBR system, the concentration of nitrate and nitrogen increased to an average value of 7.6 mg/L. Differences in the nitrate concentrations in the influent and effluent indicate that a large part of nitrogen in grey water was organically bound [16]. The average value of total Kjeldahl nitrogen in the influent was 10.6 mg/L. The removal degree of total nitrogen was at the average level of 20% and the concentration in the effluent was 8 mg/L. It should be emphasized that despite the lack of total denitrification, the concentrations of individual forms of nitrogen in SBR effluent did not exceed the maximum permissible values given that are currently in force in Poland, (Regulations of the Minister of the Environment) on the conditions to be met when discharging wastewater into water or soil [21]. At the initial stage of the grey water purification process, there was a problem with phosphorus removal. Its value exceeded several times the permissible value. It can be explained by the variable physicochemical nature of the SBR influent. The effective dephosphatation takes place when the influent contains an easily biodegradable COD fraction and the COD/BOD_5 ratio is 2 [26,27]. The second factor intensifying a release of phosphate under anaerobic conditions is the constant delivery of easily biodegradable organic compounds, e.g., volatile fatty acids and their salts. Their occurrence in grey water is likely to be changeable. Since grey water was collected in a holding tank over a longer time, it could undergo an acid fermentation during which volatile fatty acids may be formed. It was found that in the second week of the process, the degree of $P\text{-}PO_4^{3-}$ removal increased along with the increase of COD concentration of the inflowing grey water. In the following weeks of operation of the SBR reactor, the phosphate phosphorus concentration ranged from 1.1 to 2.6 $mgP\text{-}PO_4^{3-}/L$. Despite the increase in efficiency of phosphorus removal, its permissible concentration specified in the Regulation of the Minister of the Environment [21] (P_{tot} = 2 mg/L) was still exceeded. In order to improve the effectiveness of phosphate removal, the cycle of operation of SBR should be modified by means of changes in the duration of aerobic–anaerobic phases [28].

3.1.3. Removal of Color and Turbidity

Figures 4 and 5 show the treatment efficiency of the SBR system with respect to the changes in the color and the turbidity. From these results, it is clear that over the whole experiment, the efficiency of the SBR system in terms of color and turbidity was at a constant high level. During 22 days of continuous SBR system operation, the color of grey water decreased from an average value of 201 mgPt/L to 44 mgPt/L, which corresponds to an average removal of 78%. Meanwhile, the average grey water turbidity decreased from 28 FTU to 4 FTU corresponds to an average removal of 84%.

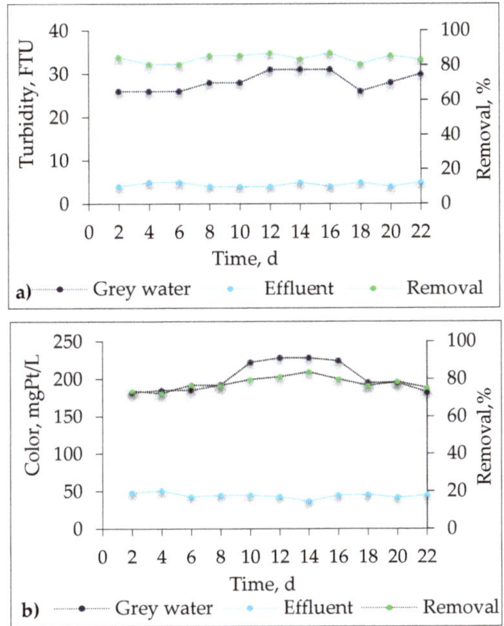

Figure 4. The removal performance of the SBR system with respect to the changes in the color and the turbidity.

Figure 5. Changes in the color and the turbidity of grey water under treatment with SBR followed by ultrafiltration.

3.2. SBR Effluent Treatment by Ultrafiltration

The treatment efficiency of SBR effluent in an ultrafiltration unit with different membranes is presented in Figure 6 and Table S3 (Supplementary file). As expected, for ultrafiltration, removal of organics expressed by the color, absorbance, COD, and TOC was very high. More specifically, the color, absorbance$_{UV254}$, COD, and TOC were reduced maximally by 73%, 91%, 84%, and 91%, respectively. Slightly lower elimination was observed for inorganic quality parameters. Conductivity, nitrate, phosphate, ammonium, and total nitrogen were reduced by 61%, 64%, 79%, 85%, and 59%. It is important to emphasize that negative charge of membranes at pH 7–8 (as seen in Figure 8) played an important role in the reduction of conductivity, phosphate, and nitrate ions in SBR effluent. The literature describes that negatively charged ultrafiltration membranes reject phosphate ions by 87%, as an effect of electrostatic repulsion [29]. Another important influencing factor on the rejection of ions and organics in the ultrafiltration unit treating wastewater/surface water can be hydrophobic/hydrophilic interactions between organics and ions in feed water [30]. For example, Shang et al. reported that high phosphate removal was attributed to the adsorption of phosphate on biopolymers in effluent and removal with these biopolymers [31].

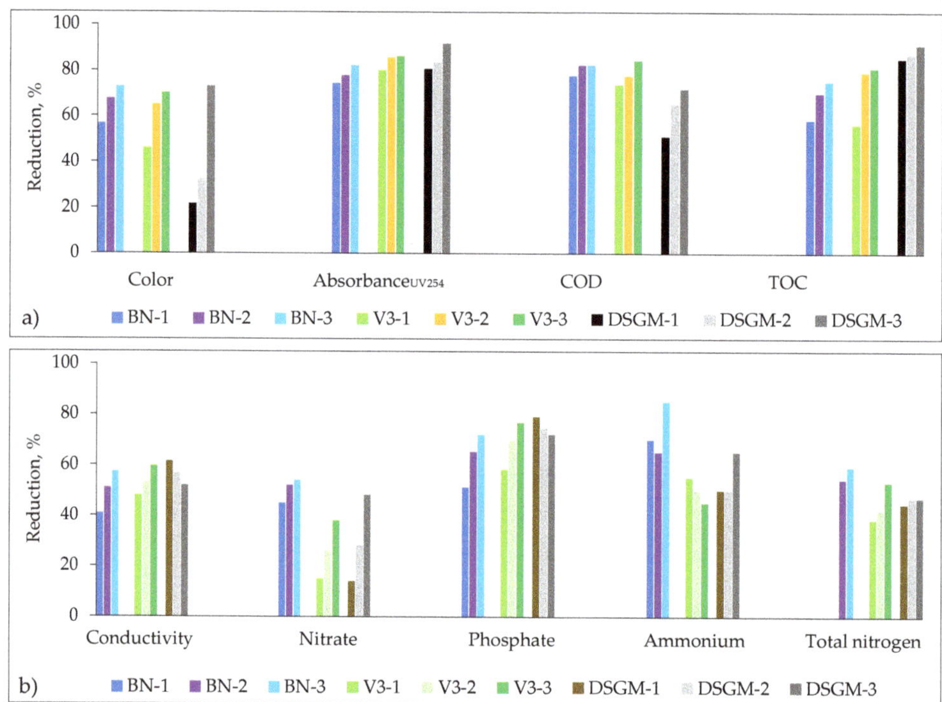

Figure 6. Reduction values of organic (**a**) and inorganic (**b**) parameters during the ultrafiltration of SBR effluent. Each bar corresponds to the filtration cycle 1–3.

Surprisingly, membrane DSGM with the lower value of MWCO provided permeate quality values very similar to the values for BN and V3 membranes with higher MWCO. This brings an important finding that the treatment efficiency was more affected by operational conditions of cross-flow ultrafiltration than by membrane properties. When good mixing and turbulent flow along the membrane surface are guaranteed, as it is in cross-flow, other factors affecting permeability and selectivity are less important [32]. Feed nature could also play an important role. From the quality parameters of SBR effluent, we can assume that SBR effluent did not contain a significant portion of low molecular weight organics that needed to be removed. Another reason could be the fouling layer that supported the retention of pollutants by BN and V3 membranes. Jerman et al. found that the cake or gel layer acts as an additional membrane in ultrafiltration and improves retention by up to 40% [33].

It was also found that treatment efficiency increased within cycles 1–3 for all membranes (Figure 6). It can be related to the formation of the cake layer on the membrane surface of BN and V3. This corresponds well with the permeability loss observed for these membranes. Many authors suggest that the cake layer decreases membrane permeability by the reduction of effective pore size and improves the retention of organics and ions [34–36]. However, in the case of the DSGM membrane, there was no fouling so the cake layer was probably not created. The reason for increasing retention over time for DSGM can be related to the concentration effect of feed components within cycles

1–3. In the literature, the influence of the feed concentration on both ions and organics removal can be found [37,38]. Muthumareeswaran et al. have found that a change of the concentration of feed components in the multicomponent system affects their retention due to the ionic interaction between the feed components (Columbic interaction), molar volume, and interaction between ions and membrane surface charge [39].

Importantly, physical, chemical, and microbiological specifications (smell, color, turbidity, TOC, ammonium, nitrate, chloride, conductivity, hardness, *E. coli*, and enterococci) of the permeate correspond well with typical tap water values of basic quality parameters [40].

3.3. Microbiological Quality of Grey Water

Pathogens, such as *Escherichia coli* and Enterococci, have been identified in grey water (Table 3). The concentration of these bacteria was above 100 CFU per 100 mL. During the biological purification process in the SBR system, the presence of *Escherichia coli* did not change, while Enterococci decreased to an average of 3 CFU per 100 mL of sample. Then, ultrafiltration for each UF membranes resulted in the complete removal of the determined pathogens [41]. High pathogen removal from the *E. coli* group was also noted in the SMBR system from Khalid Bani-Melhem et al. [3].

Table 3. Microbiological quality of grey water.

Parameter	Unit	Grey Water	SBR Effluent	UF Permeate of BN/V3/DSGM
Escherichia coli	CFU/100 mL	>100	>100	0
Enterococci	CFU/100 mL	>100	3	0

3.4. Membrane Permeability and Fouling Behavior in Ultrafiltration

Hydraulic performance of membranes was evaluated by permeability loss as a function of time. As seen in Figure 7, the permeability decreased gradually for BN and V3 membranes, while for DSGM, permeability was constant along the ultrafiltration. In other words, fouling was not observed for the DSGM membrane. This finding is very important in the context of membrane lifetime and reducing operational cost. Similarly, Acero et al. reported that UF membranes (with higher MWCO) revealed higher fouling than NF membranes (with lower MWCO) [42]. It can be explained by different surface properties of given membranes, such as hydrophilicity/hydrophobicity and surface charge (zeta potential curve and isoelectric point). The contact angle and isoelectric points of clean and fouled membranes are listed in Table 4. Zeta potential curves for clean and fouled membranes are presented in Figure 8. Owing to the lowest contact angle and strong negative surface charge, the DSGM membrane had the best antifouling properties. On the contrary, the surface of BN and V3 membranes was much more hydrophobic, and thus more prone to adsorb the feed components. Some authors reported a high importance of membrane hydrophilicity and negative charge for fouling mitigation in ultrafiltration [43–45] treating water or wastewater.

In order to investigate this further, the hydraulically reversible and hydraulically irreversible resistances were calculated. As seen in Figure 9, the total fouling (sum of reversible and irreversible resistances) was higher for the V3 membrane. However, considering the proportion of reversible and irreversible resistances for these membranes, it is clear that fouling behavior for these membranes exhibited the opposite trend. More specifically, a significant proportion of the increase in resistance for the V3 membrane was hydraulically reversible, while hydraulically irreversible for the BN membrane. Reversibility/irreversibility of fouling depends on the surface properties of membranes. It is in good correspondence with changes of course of zeta potential curves and shift in isoelectric point for clean and fouled membranes. Membrane V3 with the lower negative charge was more sensitive to fouling overall than the BN membrane. However, for the BN membrane, we observed the higher shift in isoelectric point (from 3.01 to 3.55) than for V3 (from 5.1 to 5.25) indicating a persistent change in BN membrane properties caused by irreversible fouling.

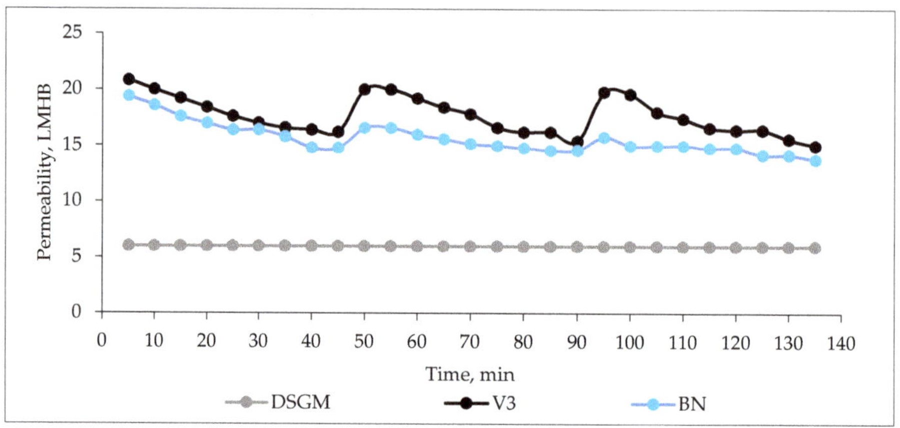

Figure 7. Permeability loss as a function of time of UF for different types of membranes. Ultrafiltration with DSGM membrane was performed without forward flushing due to constant permeability.

Table 4. Contact angle and isoelectric point for clean and fouled membranes.

Membrane	Contact Angle		Isoelectric Point	
	Clean Membrane	Fouled Membrane	Clean Membrane	Fouled Membrane
DSGM	45.5 ± 3.2	46.25 ± 2.8	-	-
V3	71.5 ± 1.9	73.5 ± 3.8	5.05	5.25
BN	75.0 ± 1.4	78.6 ± 1.2	3.01	3.55

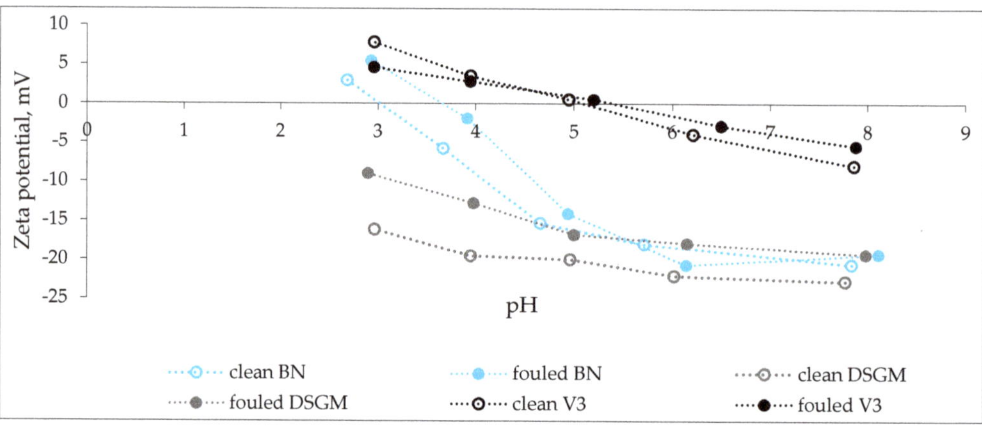

Figure 8. Zeta potential vs pH for clean and fouled membranes.

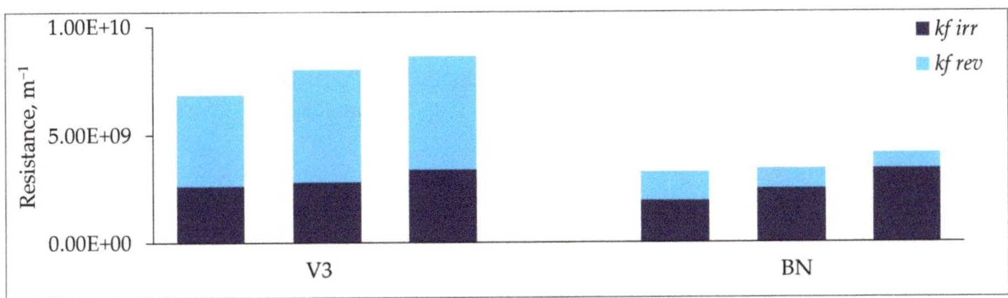

Figure 9. Irreversible and reversible fouling resistances of ultrafiltration for all membranes. Each bar corresponds to the filtration cycle 1–3. *kf irr* and *kf rev* are irreversible and reversible resistances, respectively.

4. Conclusions

This study proved a great performance of SBR followed by ultrafiltration to obtain high-quality water for non-potable purposes from real grey water. Purified grey water fulfilled criteria for wastewater discharged into water or soil as well as physical, chemical, and microbiological requirements for tap water. It is a good starting point to find new reuse application of reclaimed grey water.

The following conclusions can be reached from the experimental results.

1. The SBR system ensured the complete removal of nonionic surfactants and 97% removal of anionic surfactants from grey water.
2. Removal of phosphate in SBR was, on average, 42%, but its concentration still exceeded acceptable values. It could probably be related to the variable concentration of raw grey water. It was stated that in order to improve the phosphorus reduction, in subsequent studies, the cycle of operation of SBR should be modified by means of changes in the duration of aerobic–anaerobic phases.
3. In ultrafiltration treatment, there was an efficiency increase in filtration time due to the formation of a cake layer or concentration effect.
4. Treatment efficiency in ultrafiltration was mainly influenced by the cross-flow filtration mode and nature of feed water.
5. Membrane properties played an important role in fouling behavior, but did not greatly affect treatment efficiency.
6. Membrane DSGM did not show fouling in permeate flux monitoring. It is a consequence of the initial properties of this membrane such as the lowest contact angle and the most negative zeta potential when compared with the BN and V3 membranes, for which fouling was observed.

Supplementary Materials: The following is available online at http://www.mdpi.com/2073-4441/12/1/154/s1, Table S1: Codes of the Merck test kits, Table S2: Values of quality parameters and reduction degree for SBR system, Table S3: Values of quality parameters measured in SBR effluent and permeate samples.

Author Contributions: Conceptualization G.K. and A.M. methodology, G.K. and A.M.; investigation, G.K. and A.M.; writing—original draft preparation, G.K. and A.M., Writing—review and editing, G.K. All authors have read and agreed to the published version of the manuscript.

Funding: This research was funded by the Polish Ministry of Science and Higher Education.

Conflicts of Interest: The author declares no conflicts of interest. The funders had no role in the design of the study; in the collection, analyses, or interpretation of data; in the writing of the manuscript, or in the decision to publish the results.

References

1. Use of Freshwater Resources. Available online: https://www.eea.europa.eu/data-and-maps/indicators/use-of-freshwater-resources-2/assessment-3 (accessed on 23 December 2019).

2. Xin, D.; Saige, W.; Bin, C. The blue, green and grey water consumption for crop production in Heilongjiang. *Energy Procedia* **2019**, *158*, 3908–3914.

3. Bani-Melhem, K.; Al-Qodah, Z.; Al-Shannag, M.; Qasaimeh, A.; Qtaishat, M.R.; Alkasrawi, M. On the performance of real grey water treatment using a submerged, Membrane bioreactor system. *J. Membr. Sci.* **2015**, *476*, 40–49. [CrossRef]

4. Martínez-Alcalá, I.; Pellicer-Martínez, F.; Fernández-López, C. Pharmaceutical grey water footprint: Accounting, influence of wastewater treatment plants and implications of the reuse. *Water Res.* **2018**, *135*, 278–287. [CrossRef] [PubMed]

5. Galvis, A.; Zambrano, D.A.; Van der Steen, N.P.; Gijzen, H.J. Evaluation of pollution prevention options in the municipal water cycle. *J. Clean. Prod.* **2014**, *66*, 599–609. [CrossRef]

6. Prajapati, B.; Jensen, M.B.; Jørgensen, N.O.G.; Petersen, N.B. Grey water treatment in stacked multi-layer reactors with passive aeration and particle trapping. *Water Res.* **2019**, *161*, 181–190. [CrossRef]

7. WHO. Guidelines for the Safe Use of Wastewater, Excreta and Greywater, Excreta and Greywater Use in Agriculture. 2006. Available online: https://apps.who.int/iris/bitstream/handle/10665/78265/9241546824_eng.pdf;jsessionid=47807504FED628D9A9BB71654F04A9E7?sequence=1 (accessed on 23 December 2019).

8. Li, F.; Wichmann, K.; Otterpohl, R. Review of the technological approaches for grey water treatment and reuses. *Sci. Total Environ.* **2009**, *407*, 3439–3449. [CrossRef]

9. Boddu, M.V.; Paul, T.; Page, M.A.; Byl, C.; Ward, L.; Ruan, J. Gray water recycle: Effect of pretreatment technologies on low pressure reverse osmosis treatment, *J. Environ. Chem. Eng.* **2016**, *4*, 4435–4443. [CrossRef]

10. Manouchehri, M.; Kargari, A. Water recovery from laundry wastewater by the cross flow microfiltration process: A strategy for water recycling in residential buildings. *J. Clean. Prod.* **2017**, *168*, 227–238. [CrossRef]

11. Ding, A.; Liang, H.; Li, G.; Derlon, N.; Szivak, I.; Morgenroth, E.; Pronk, W. Impact of aeration shear stress on permeate flux and fouling layer properties in a low pressure membrane bioreactor for the treatment of grey water. *Water Res.* **2016**, *510*, 382–390. [CrossRef]

12. Mohammadi, M.J.; Takdastan, A.; Jorfi, S.; Neisi, A.; Farhadi, M.; Yari, A.R.; Dobaradaran, S.; Khaniabadi, Y.O. Electrocoagulation process to Chemical and Biological Oxygen Demand treatment from carwash grey water in Ahvaz megacity, Iran. *Data Brief* **2017**, *11*, 634–639. [CrossRef]

13. Bani- Melhem, K.; Smith, E. Grey water treatment by a continuous process of an electrocoagulation unit and a submerged membrane bioreactor system. *Chem. Eng. J.* **2012**, *198–199*, 201–210. [CrossRef]

14. Ding, A.; Liang, H.; Li, H.; Szivak, I.; Traber, J.; Pronk, W. A low energy gravity driven membrane bioreactor system for grey water treatment: Permeability and removal performance of organics. *J. Membr. Sci.* **2017**, *542*, 408–417. [CrossRef]

15. Leong, J.; Oha, K.S.; Poh, P.E.; Chong, M.N. Prospects of hybrid rainwater-greywater decentralised system for water recycling and reuse: A review. *J. Clean. Prod.* **2017**, *142*, 3014–3027. [CrossRef]

16. Adamczak, M.; Kamińska, G.; Bohdziewicz, J. Application of waste polymers as basic material for ultrafiltration membranes preparation. *Proceedings* **2019**, *16*, 14. [CrossRef]

17. Drews, A. Membrane fouling in membrane bioreactors—Characterisation, contradictions, cause and cures. *J. Membr. Sci.* **2010**, *363*, 1–28. [CrossRef]

18. Duan, L.; Jiang, W.; Song, Y.; Xia, S.; Hermanowicz, S. The characteristics of extracellular polymeric substances and soluble microbial products in moving bed biofilm reactor-membrane bioreactor. *Bioresour. Technol.* **2013**, *148*, 436–442. [CrossRef]

19. Pinto, A.C.S.; Grossi, L.; Carvalho de Melo, R.A.; Macedo de Assis, T.; Ribeiro, V.M.; Santos Amaral, M.C.; de Souza Figueiredo, K.C. Carwash wastewater treatment by micro and ultrafiltration membranes: Effects of geometry, pore size, pressure difference and feed flow rate in transport properties. *J. Water Process Eng.* **2017**, *17*, 143–148. [CrossRef]

20. Crittenden, J.C.; Trussell, R.R.; Hand, D.W.; Howe, K.J.; Tchobanoglous, J. *Water Treatment: Principles and Design*, 3rd ed.; John Wiley & Sons: Hoboken, NJ, USA, 2012.

21. Dz.U.2014.poz.1800. Directive of the Minister of the Environment from 18th of November 2014. Available online: http://prawo.sejm.gov.pl/isap.nsf/download.xsp/WDU20140001800/O/D20141800.pdf (accessed on 3 January 2020).

22. Fountoulakis, M.S.; Markakis, N.; Petousi, I.; Manios, T. Single house on-site grey water treatment using a submerged membrane bioreactor for toilet flushing. *Sci. Total Environ.* **2016**, *551–552*, 706–711. [CrossRef]

23. Guang-Guo, Y. Fate, behavior and effects of surfactants and their degradation products in the environment. *Environ. Int.* **2006**, *32*, 417–431.

24. Ivanković, T.; Hrenović, J. Surfactants in the Environment. *Arh. Hig. Rada Toksikol.* **2010**, *61*, 95–110. [CrossRef]

25. Leal, H.; Temmink, H.; Zeeman, G.; Buisman, C.J.N. Comparison of Three Systems for Biological Greywater Treatment. *Water* **2010**, *2*, 155–169. [CrossRef]

26. Świerczynska, A.; Bohdziewicz, J. Determination of the most effective operating conditions of membrane bioreactor used to industrial wastewater treatment. *Environ. Prot. Eng.* **2015**, *41*, 41–51.

27. Hocaoglua, S.M.; Atasoya, E.; Babana, A.; Orhonb, D. Modeling biodegradation characteristics of grey water in membrane bioreactor. *J. Membr. Sci.* **2013**, *429*, 139–146. [CrossRef]

28. Świerczyńska, A.; Bohdziewicz, J.; Puszczało, E. Treatment of industrial wastewater in the sequential membrane bioreactor. *Ecol. Chem. Eng. S* **2016**, *23*, 285–295. [CrossRef]

29. Shang, R.; Verliefde, A.R.D.; Hu, J.; Zeng, Z.; Lu, J.; Kemperman, A.J.B.; Deng, H.; Nijmeijer, K.; Heijman, S.G.J.; Rietveld, L.C. Tight ceramic UF membrane as RO pre-treatment: The role of electrostatic interactions on phosphate rejection. *Water Res.* **2014**, *48*, 498–507. [CrossRef] [PubMed]

30. Peeva, P.D.; Million, N.; Ulbricht, M. Factors affecting the sieving behavior of anti-fouling thin-layer cross-linked hydrogel polyethersulfone composite ultrafiltration membranes. *J. Membr. Sci.* **2012**, *390–391*, 99–112. [CrossRef]

31. Shang, R.; Verliefde, A.R.D.; Hu, J.; Heijmann, S.G.J.; Rietveld, L.C. The impact of EfOM, NOM and cations on phosphate rejection by tight ceramic ultrafiltration. *Sep. Purif. Technol.* **2014**, *132*, 289–294. [CrossRef]

32. Broeckmann, A.; Busch, J.; Wintgens, T.; Marquardt, W. Modeling of pore blocking and cake layer formation in membrane filtration for wastewater treatment. *Desalination* **2006**, *189*, 97–109. [CrossRef]

33. Jermann, D.; Pronk, W.; Boller, M.; Schafer, A.I. The role of NOM fouling for the retention of estradiol and ibuprofen during ultrafiltration. *J. Membr. Sci.* **2009**, *329*, 75–84. [CrossRef]

34. Bohdziewicz, J.; Kamińska, G.; Pawlyta, M.; Łukowiec, D. Comparison of effectiveness of advanced treatment of municipal wastewater by sorption and nanofiltration. Separate processes and integrated systems. *Environ Prot. Eng.* **2015**, *41*, 119–132.

35. García-Martín, N.; Silva, V.; Carmona, F.J.; Palacio, L.; Hernández, A.; Prádanos, P. Pore size analysis from retention of neutral solutes through nanofiltration membranes. The contribution of concentration–polarization. *Desalination* **2014**, *344*, 1–11. [CrossRef]

36. Meng, S.; Zhanf, M.; Yao, M.; Qiu, Z.; Hong, Y.; Lan, W.; Xia, H.; Jin, X. Membrane Fouling and Performance of Flat Ceramic Membranes in the Application of Drinking Water Purification. *Water* **2019**, *11*, 2606. [CrossRef]

37. Kramer, F.C.; Shang, R.; Rietveld, L.C.; Heijman, S.J.G. Influence of pH, multivalent counter ions, and membrane fouling on phosphate retention during ceramic nanofiltration. *Sep. Purif. Technol.* **2019**, *227*. [CrossRef]

38. Arsuaga, J.M.; López-Muñoz, M.J.; Aguado, J.; Sotto, A. Temperature, pH and concentration effects on retention and transport of organic pollutants across thin-film composite nanofiltration membranes. *Desalination* **2008**, *221*, 253–258. [CrossRef]

39. Muthumareeswaran, M.R.; Agarwa, G.P. Feed concentration and pH effect on arsenate and phosphate rejection via polyacrylonitrile ultrafiltration membrane. *J. Membr. Sci.* **2014**, *468*, 11–19. [CrossRef]

40. Acero, J.L.; Benitez, F.J.; Teva, F.; Leal, A.I. Retention of emerging micropollutants from UP water and a municipal secondary effluent by ultrafiltration and nanofiltration. *Chem. Eng. J.* **2010**, *163*, 264–272. [CrossRef]

41. Haas, R.; Opitz, R.; Grischek, T.; Otter, P. The AquaNES Project: Coupling Riverbank Filtration and Ultrafiltration in Drinking Water Treatment. *Water* **2019**, *11*, 18. [CrossRef]

42. Świerczyńska, A.; Bohdziewicz, J.; Kamińska, G.; Wojciechowski, K. Influence of the type of membrane-forming polymer on the membrane fouling. *Environ. Prot. Eng.* **2016**, *42*, 197–210.

43. Ghiasi, S.; Behboudi, A.; Mohammadi, T.; Khanlari, S. Effect of surface charge and roughness on ultrafiltration membranes performance and polyelectrolyte nanofiltration layer assembly, Colloids and Surfaces A. *Physicochem. Eng. Asp.* **2019**, *580*. [CrossRef]

44. Kumari, P.; Modi, A.; Bellare, J. Enhanced flux and antifouling property on municipal wastewater of polyethersulfone hollow fiber membranes by embedding carboxylated multi-walled carbon nanotubes and a vitamin E derivative. *Sep. Purif. Technol.* **2020**, *235*. [CrossRef]
45. Regulation of the Ministry of Health on the Quality of Water Intended for Human Consumption. Available online: https://www.igwp.org.pl/images/artykuly/prawo/Jako%C5%9B%C4% 87-wody-przeznaczonej-do-spo%C5%BCycia-przez-ludzi.pdf (accessed on 23 December 2019).

 *water*

Article

Olive Mill Wastewater: From a Pollutant to Green Fuels, Agricultural Water Source, and Bio-Fertilizer. Part 2: Water Recovery

Patrick Dutournié [1], Mejdi Jeguirim [1,*], Besma Khiari [2], Mary-Lorène Goddard [3,4] and Salah Jellali [5]

[1]   CNRS ISM2 UMR 7361, Université de Haute Alsace, Université de Strasbourg, 68093 Mulhouse, France; Patrick.dutournie@uha.fr
[2]   National School of Engineers of Carthage, 45 rue des entrepreneurs, Charguia 2–1002, 2035 Tunis, Tunisia; besmakhiari@yahoo.com
[3]   CNRS, LIMA UMR 7042, Université de Haute-Alsace, Université de Strasbourg, 68093 Mulhouse, France; mary-lorene.goddard@uha.fr
[4]   LVBE, EA 3991, Université de Haute-Alsace, 68008 Colmar, France
[5]   Wastewaters and Environment Laboratory, Water research and Technologies Centre, 8020 Soliman, Tunisia; salah.jallali@certe.rnrt.tn
*   Correspondence: mejdi.jeguirim@uha.fr; Tel.: +33-389-6729

Received: 20 March 2019; Accepted: 10 April 2019; Published: 13 April 2019

Abstract: Water shortage is a very concerning issue in the Mediterranean region, menacing the viability of the agriculture sector and in some countries, population wellbeing. At the same time, liquid effluent volumes generated from agro-food industries in general and olive oil industry in particular, are quite huge. Thus, the main aim of this work is to suggest a sustainable solution for the management of olive mill wastewaters (OMWW) with possible reuse in irrigation. This work is a part of a series of papers valorizing all the outputs of a three-phase system of oil mills. It deals with recovery, by condensation, of water from both OMWW and OMWW-impregnated biomasses (sawdust and wood chips), during a convective drying operation (air velocity: 1 m/s and air temperature: 50 °C). The experimental results showed that the water yield recovery reaches about 95%. The condensate waters have low electrical conductivity and salinities but also acidic pH values and slightly high chemical oxygen demand (COD) values. However, they could be returned suitable for reuse in agriculture after additional low-cost treatment.

Keywords: OMWW; drying; water recovery; water characterization; sustainable development

1. Introduction

Water scarcity in the Mediterranean region is becoming a growing concern, menacing the viability of agriculture, which represents an important economic sector in many countries [1]. This water shortage is accentuated in the last decades due to the undeniable climate change causing recurrent long periods of drought. The efficient wastewaters treatment and their controlled reuse in irrigation can contribute to water saving and potentially address this water shortage issue [2,3].

Generally, treated municipal wastewater is the main recyclable source for irrigation. This treatment and reuse process contributes strongly to the potable water resources conservation and reduces the wastewater's negative impacts related to the effluents release into the environment [4]. However, the wastewater chemical composition needs to be controlled due to the presence of a variety of pollutants including heavy metals, organic compounds, bacteria, etc. [5,6]. These contaminants may generate problems for agricultural production affecting crop quantity and quality [7,8]. Therefore, the interest turned towards agro-industrial wastewaters, because of the large amounts produced and their relative high contents in nutrients [9].

Still, in the Mediterranean region, the olive oil industry generates large quantities of olive byproducts that should be appropriately managed [10]. Different kinds of extraction processes are commonly used. The three-phase system, mainly used in Tunisia, Greece, and Italy, generates a solid residue (named olive mill solid waste (OMSW)) and a liquid effluent (olive mill wastewater (OMWW)). Approximately, 30 million tons of OMWW are generated each year in the Mediterranean region [11]. The OMWW, which is slightly acidic, contains a large fraction of water (around 80%) and high contents of soluble organic compounds and salts [12]. These wastes are often discharged and stored in natural open-air basins (Figure 1a). Indeed, the OMWW surface rapidly dries (as a result of the heat supplied by air flow and sun heating) and a crust (with a plastic consistency) rapidly covers the surface (Figure 1b–d). Such crust inhibits the mass and heat transfers and, therefore, water evaporation. In these conditions, the aqueous phase could diffuse and percolate in the subsoil, leading to soils infertility and groundwater pollution. Furthermore, it is the origin of bad smells and the development of several varieties of mosquitoes.

(a)   (b)

(c)   (d)

Figure 1. Photos of olive mill wastewaters (OMWW) evolution in deposit basins in the region of Mahdia (Tunisia) at different drying stage.

Sustainable OMWW management has been pointed out as an urgent challenge in order to tackle the disadvantages cited above [11–13]. More than 20 applicable procedures are mentioned in scientific publications, including elemental operations such as flocculation, ultrafiltration, and chemical treatments or combined operations such as centrifugation–ultrafiltration. These techniques are generally tested in a laboratory or in a pilot plant, without posterior industrial projections. Some treatment and valorization systems presenting some degree of applicability are summarized in Table 1.

Table 1. Main applied technologies for OMWW treatment.

Technology	Description	Advantages	Drawbacks
Land spreading	OMWW is used as amendment (30–80 m^3 ha^{-1} year^{-1})	- Low cost operation - Soil fertility and crops growth	• Effect on soil is controversial • Possibility of groundwater contamination through infiltration
Composting	Aerobic breakdown of organic content of OMWW	- soil fertility and crops growth	• Location far from urban or transit zones (bad smell) • Possibility of Groundwater contamination by drainage water
Thermal treatment	OMWW mixing with dry olive husk wastes	- Wastes reduction and energy recovery	• High cost of drying • High investment and operating costs
Membrane filtration	Separation of the solid fraction using multilayered filters	- Treatment of high flowrates - Water recovery and solid subsequent reuse	• High energy consumption • High risk of filters clogging
Biological treatment	Biological degradation of the organic matter in anaerobic digesters	- Low energy consumption	• Expensive plantings and machinery • Production of high amounts of activated sludge
Coagulation-flocculation	Aggregation of the suspended matter through polyers addition	- Water recovery	• Expensive plantings and machinery • Production of high amounts of activated sludge
Electrocoagulation	Imposition a current between two electrodes immersed in an electrolyte to produce a coagulant in solution	- Water recovery - No necessity of using expensive machinery, no use of moving parts	• High energy consumption • Production of high amounts of sludge containing heavy metals
Adsorption	Elimination of organic and mineral content by using porous media	- Low cost, availability, and flexibility - Water recovery	• Pollution transfer from liquid to solid phase • Limited treatment yield

The agronomic use of OMWW through its spreading on soils, directly or following a composting process, has several drawbacks, including the modification of soils' physical, chemical, and biological properties. On the contrary, the thermal treatment could be an interesting solution since it permits the recovery of biofuels (for energy production) and biochars (for agronomic and environmental purposes). However, the high moisture content is not economically favorable for the direct OMWW thermochemical conversion [14,15]. In order to face such constraint, the OMWW impregnation on lignocellulosic biomass was proposed to recuperate the organic matter followed by agropellets production [16]. Nevertheless, the pellets' combustion in domestic boilers has some drawbacks, like particulate matters emissions and ash accumulation, due to the high mineral contents in OMWW [17].

Recently, an environmentally friendly strategy for the OMWW valorization, including different procedures, was implemented [18]. The first step was the impregnation of wood sawdust with OMWW, leading to the recovery of the OMWW's organic and nutrient compounds. The second step of drying showed that the impregnation stage led, not only to the oily fraction retention, but also to the acceleration of water evaporation kinetic. The third step consisted of the pyrolysis of these impregnated samples, which led to high yields of chars with attractive nutrients concentrations, especially potassium and phosphorus. The last step concerned the application of the obtained chars as bio-fertilizers. This operation confirmed a beneficial effect on the plant growth [18].

More precisely, and as far as the drying step is concerned, the experiments were carried out in a convective dryer under different operating conditions (velocity, temperature). The kinetic results confirmed that the drying of the impregnated samples was faster than that of OMWW samples. But more importantly, such procedure could probably allow an ecologic recovery of water from OMWW [19]. In this vein, this paper aims to examine the water recovery from the drying step and its characterization for a possible reuse in irrigation. Indeed, water management will be increasingly strategic, especially in the Mediterranean basin. Very few works in the literature deal with the recovery of water for an agricultural purpose after a drying operation [20]. These studies are focused on the recovering of added-value chemicals in water, for example, essential oils extraction [21].

2. Materials and Methods

2.1. Samples Preparation

Olive mill wastewater (OMWW) used in this study was provided by an olive mill located in the city of Touta, North East of Tunisia. This olive mill plant uses the three-phase extraction system for olive oil production. Sawdust was provided by Nollinger sawmill located in Illfurth (France). Wood chips were supplied by the firewood supplier Farmingroad located in Reguisheim (France).

Five samples were prepared during this investigation in order to examine the water recovery process and analyze its quality for the possible reuse in agriculture. The raw OMWW sample (moisture content: 91% wet basis, wb.) was used for the impregnation procedure. This sample was also used without impregnation for the water recovery tests. The impregnation samples were prepared according to the procedure applied in previous investigations [16,17]. Impregnated sawdust (IS) was prepared by mixing 2 kg of OMWW with 0.5 kg of sawdust (moisture content: 24% wb.) for 24 h. The initial moisture content of the IS before drying was 78% wb. In a similar way, impregnated wood chips (IWC) were prepared by mixing 1.5 kg of OMWW with 0.5 kg of wood chips (moisture content: 24% wb.). The initial moisture of IWC before the drying test was 74% wb.

Two complementary samples (blank tests) were required in order to identify the effect of the OMWW impregnation on the water quality. Hence, humidified sawdust (HS) was prepared by emerging 0.5 kg of sawdust in 2 L deionized water respecting the same procedure as for OMWW impregnation. Humidified wood chips (HWC) were prepared by emerging 0.5 kg of wood chips in 1.5 L deionized water.

2.2. Experimental Drying Tests

Experimental drying tests were performed using a laboratory convective dryer (Figure 2) previously described in details [19].

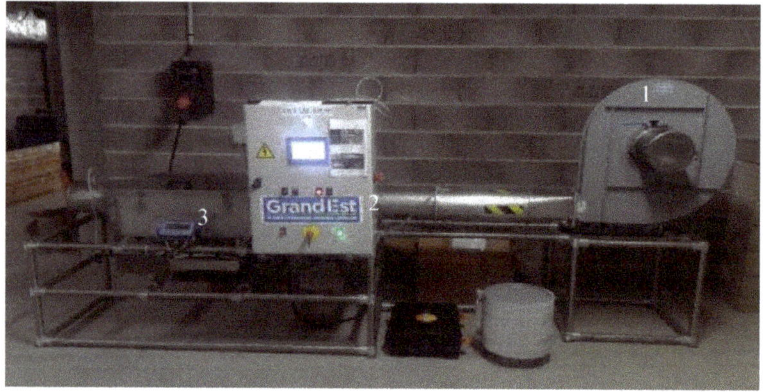

Figure 2. Experimental dryer.

During the drying tests, samples were placed in an open Plexiglas box (19 × 13 × 3 cm). The level of wet samples is accurately adjusted to the upper rim of the box. The mass of the studied sample was recorded in order to measure the moisture content progress versus time. A fan (1) introduces air in the dryer with an air velocity of 1 m/s. The fan is settled for having an air velocity directly onto the material surface. The air heating is achieved by electrical resistances (2) at 50 °C. The air temperature in the drying chamber is controlled by a proportional–integral–derivative (PID) control loop. The operating drying conditions (1 m/s and T = 50 °C) were selected from previous results [19] in order to have a good compromise between the drying time and the energy consumption. Temperature, humidity, and flow rate are continuously measured by several sensors; 1 for fluid velocity in the duct, 3 humidity sensors (room air, hot air at the drying chamber inlet, and at the dryer outlet) and 7 thermocouples (room air, hot air, 4 in the drying chamber, and 1 at the dryer outlet).

From mass measurements, the moisture content X is given by:

$$X(t) = \frac{m(t) - m_f}{m(t)} \tag{1}$$

where m(t) is the sample weight at time t and m_f is the sample weight at the end of the drying.

2.3. Water Condensation and Recovery

In order to examine the water recovery from the impregnated samples, an accessory material was implemented at the dryer outlet. In particular, a boiler condenser (Figure 3) was used for the water condensation at the outlet of the dryer. For this purpose, the moist air flow is introduced in the boiler exchanger. The latter is cooled by an antifreeze fluid circulating into the aluminum jacket.

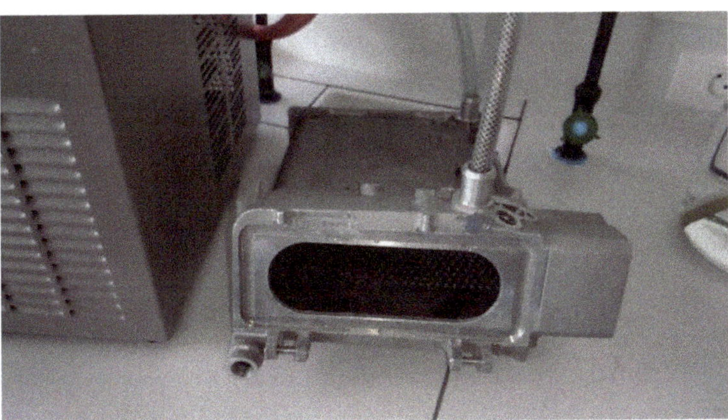

Figure 3. Boiler condenser used for the water recovery experiments.

2.4. Condensed Water Characterization

The condensed water recovered after the drying operation of raw OMWW, IS, IWC, HS, and HWC was physico-chemically characterized by using different analytical techniques in order to assess their potential reuse for irrigation in agriculture. This characterization concerned the assessment of: (i) Their pH by using a HI 2211 (HANNA Instrument, Lingolsheim, France) apparatus, (ii) their electrical conductivity (salinity) through a PC 5000 L (VWR, Radnor, PA, USA) device, (iii) their chemical oxygen demand (COD) via the open reflux using dichromate titrimetric method [22], and (iv) their contents in inorganic anions and cations through ion chromatography analyzer (Metrohm, Herisau, Switzerland).

2.5. Gas Chromatography–Mass Spectrometry

2.5.1. Chemicals and Reagents

Reagents were purchased from Sigma–Aldrich (St Quentin Fallavier, France) for methoxyamine hydrochloride and Alfa Aesar (by Thermo Fisher, Karlsruhe, Germany) for anhydrous pyridine and MSTFA). LC-MS grade water was purchased from Fisher Scientific (Illkirch, France).

2.5.2. Derivatization

A volume of 30 μL of each sample was lyophilized and then re-dissolved in 20 μL of 30 mg mL^{-1} methoxyamine hydrochloride in anhydrous pyridine and derivatized at 37 °C for 120 min with mixing at 600 rpm. The samples were incubated for 30 min with mixing at 600 rpm after addition of both 80 μL of N-methyl-N-(trimethylsilyl)trifluoroacetamide (MSTFA). Each derivatized sample was allowed to rest for 30 min prior to injection.

2.5.3. GC–MS Instrument Conditions

Samples (1 μL) were injected into a GC–EI–MS system comprising an AOC-20i Auto-injector, a Shimadzu GC-2010 gas chromatograph, and a Shimadzu GCMS-QP2010 mass spectrometer (Shimadzu, Kyoto, Japan) with an electron impact (EI) ion source and quadrupole analyzer. The GC was operated in constant linear velocity mode (40.3 cm s^{-1}) with helium as the carrier gas and using phenoxyacetic acid as a standard for retention time locking of the method. The MS was adjusted using perfluorotributylamine (PFTBA). An SGE Analytical Science BPX5 column (25 m long, 0.15 mm inner diameter, 0.25 μm film thickness) was used. The injection temperature in splitless mode was set at 310 °C, the MS transfer line at 330 °C, and the ion source adjusted to 200 °C. Helium was used as the carrier gas at a flow rate of 0.97 mL min^{-1}. The following temperature program was used; injection at 110 °C, hold for 4 min, followed by a 10.5 °C min^{-1} oven temperature, ramp to 155 °C, then 11.5 °C min^{-1} oven temperature, ramp to 350 °C, and a final 6 min heating at 350 °C.

2.5.4. Chemical Identification

Gas Chromatography–Mass Spectrometry (GC-MS) results are analyzed from the National Institute of Standards and Technology (NIST) 05 database and identified chemicals are validated by comparison with injection of commercial standards.

3. Results

3.1. Kinetic Study

Figure 4 shows the evolution of the drying in time of the OMWW, IS, and IWC in terms of moisture content X versus time.

One can see that for the three samples, the moisture ratio decreased continuously with the drying time, with different constant rate drying period. This is in accordance with other works dealing with the drying of OMWW [23] and of other biomasses impregnated with OWWW [19,23].

During the first phase of drying, the slope of the moisture curve was very dependent on the sample used. Indeed, the highest one was observed for the IWC, followed by IS, and OMWW. As this stage corresponded to the departure of free or/and weakly bound water, one can attribute the difference to the mixture density and its porosity. Indeed, there were more voids within the impregnated wood chips (IWC), which led to a better air infiltration, whereas the impregnated sawdust (IS) was more compact and formed a paste-like mixture. As for the OMWW, the oily layer transformed into a crust at the top of the liquid, which may explain the slow evaporation rate and why the first stage ended earlier (16 h) compared to IWC (19.5 h) and IS (33.8 h).

In the second drying stage (drying stage separation shown by a dotted line in Figure 3), this tendency was inversed and the slopes were in the following order: OMWW > IS > IWC.

The acceleration in OMWW drying, during the second phase, could be the result of the cracking of the crust layer (white patch shown in Figure 5, circled in red), due to the evaporated water under the crust forcing a way out on one hand and to the mechanical action of the convective air on the other hand. The resulting fissures then allow water to escape, and therefore the drying rate is increased.

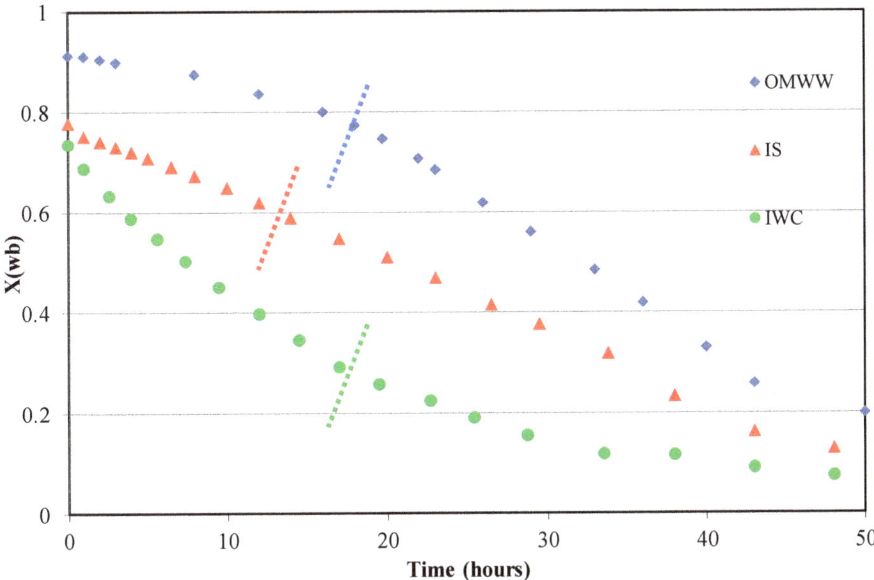

Figure 4. Moisture content profile during the convective drying for raw olive mill wastewater (OMWW), impregnated sawdust (IS), and impregnated wood chips (IWC).

The reason the IS drying rate was higher than that of IWC during the second stage may be related to the impregnation procedure and the material nature. Indeed, as sawdust is a finer and thinner material than the wood chips, its mixture with the oily liquid might have promoted the formation of stronger chemical bonds between the solid matrix and the liquid phase. The phenomenon of water and oil retention on the biomass fibers is probably the reason their slopes were inferior.

For the two impregnated solids, the water loss rate was less in the second stage, compared to the first one. Following the decrease in the concentration gradient between the samples and the surrounding air, mass flow was limited within the solid samples. This is due to bound water, which is more difficult to remove. In fact, the drying process is limited by a diffusion phase within the solid matrix. The vaporization front gradually moves towards the interior of the material. As water vapor has a longer way to cross, the surface pressure decreases. The difference between the latter and the vapor pressure of the surrounding air also decrease, slowing down the exchange and consequently the rate of drying [24].

Finally, it is important to point out that the drying of the impregnated biomasses was faster than that of the OMWW sample ($\times$ 2 for IS and $\times$ 3.5 for IWC) during only the first stage of drying. Such a conclusion was reached for impregnated olive cake and impregnated sawdust (two times faster than OMWW for both materials) but for the whole operation of drying [19]. This may be linked to the fact that in the mentioned paper, the layer to dry was thin (3 mm) whereas it was 3 cm in the present study. During the second stage, the drying was lower ($\times$ 1/2 for IS and $\times$ 1/3.5 for IWC) compared to OMWW, which explains why the amount of water evaporated per hour was substantially the same for all the samples.

Figure 5. IWC, OMWW, and IS before and after drying.

3.2. Water Recovery

Figure 6 shows the ratio of recovered water by the initial water during the drying operation for OMWW, IS, and IWC. The final ratios of condensed water are interesting since they were around 80% for OMWW and IWC and as high as 95% for IS. In a context of water shortage, such yields are very promising for further subsequent use.

The impregnation process not only shortened the drying time during the first step, but also increased the contained water recovery ratio by 18% to 35% during the whole condensation operation. Moreover, dried IS samples have the shape of bricks (Figure 7) with good mechanical properties on one hand and with increased high heating values (18 MJ/kg) compared to non-impregnated ones (16.4 MJ/kg) on the other hand. These bricks are homogenous, well-bonded, and can be easily removed from the container, contrarily to the dried OMWW where the oily and plastic phase makes it stick to the box and hard to remove. All these aspects make dried IS easy to handle, transport, and store and, suitable for direct use as bio-fuels or for a densification process by briquetting or pelletizing.

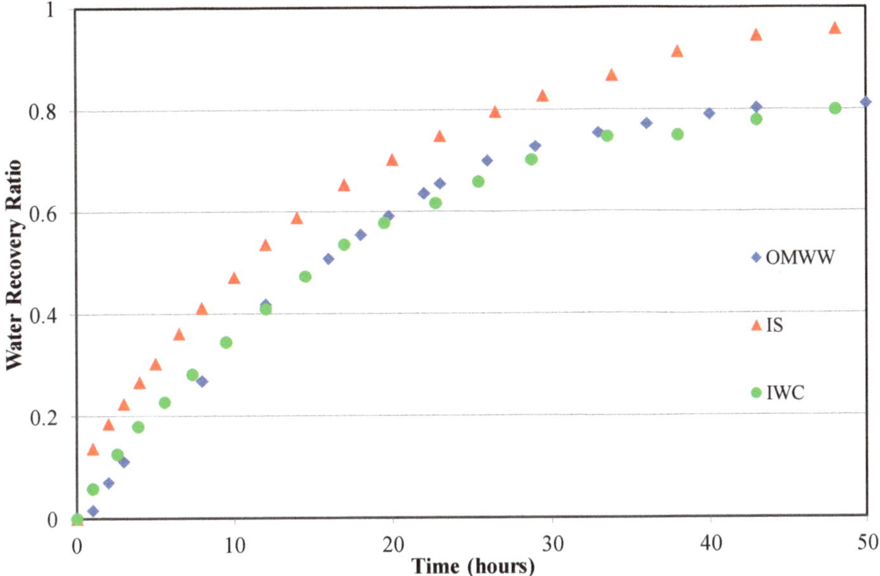

Figure 6. Water recovered ratio during drying for OMWW, IS, and IWC.

Figure 7. Impregnated sawdust at the end of the drying operation.

3.3. Recovered Water Characterization

Figure 8 shows the aspect of recuperated condensed water (right) and of the raw OMWW (left). Despite the clearness of the obtained waters, their characterization is essential to ensure their compliance with water irrigation standards.

Figure 8. Aspect of OMWW (left) and recovered-condensed water from raw OMWW (right).

As mentioned in Section 2.4, several parameters were measured for this purpose. Table 2 gives these main physico-chemical properties of the raw OMWW and the recovered water from the drying operation of OMWW, IS, IWC, HS, and HWC, in comparison with the Tunisian standard for both discharging in water bodies and reuse in agriculture (norms NT 4106.002, NT 106.003 and decree-law N°89-1047 of 28 July 1989).

Table 2. Main parameters of raw OMWW and allowable concentrations (standards).

Parameter	Raw OMWW	OMWW	IS	IWC	HS	HWC	Discharging Norm	Reuse in Agriculture Norm
pH	4.8	3.5	3.9	3.8	3.8	3.7	6.5–8.5	6.5–8.5
EC (mS/cm)	9.73	0.25	0.23	0.27	0.05	0.02	5.00	7.00
COD (g/L)	100.00	8.38	2.10	8.40	0.10	0.04	0.13	0.09
Na^+ (mg/L)	2465.0	2.0	3.3	1.2	3.3	2.6	700	-
K^+ (mg/L)	5046.5	5.0	1.9	2.0	2.1	4.5	50	-
Mg^{2+} (mg/L)	1462.5	0.4	1.1	0.4	0.8	1.0	300	-
Ca^{2+} (mg/L)	1907.8	1.7	4.3	1.5	3.0	3.9	500	-
P_{total} (mg/L)	832.1	0.7	1.6	1.4	24.8	1.9	2	-
NO_3^- (mg/L)	1056.8	1.5	1.6	0.4	3.0	5.5	600	-
Cl^- (mg/L)	1912.9	4.8	4.3	2.3	8.5	2.6	700	2000.0

According to Table 2, all the raw OMWW physico-chemical characteristics were much higher than the discharging and reuse requirements. For instance, the OMMW' COD contents and phosphorus concentrations were about 769- and 416-fold higher than the related fixed maximum concentrations, respectively.

At the same time, the drying operation significantly decreased the pH of the recovered water solution compared to the raw OMWW (4.77). Indeed, the pH values of the recovered solutions varied between 3.9 for IS and 3.5 for OMWW. These values are again very low compared to the fixed norms of discharging and reuse (Table 2). However, their adjustment to these norms could be easily performed by filtrating these recovered solutions through low-cost alkaline wastes such as seashell [25] or powdered marble [26]. Furthermore, the drying procedure also significantly decreased the electrical conductivity of the raw OMWW from 9.73 mS cm^{-1} to about 0.25 mS cm^{-1} for the recovered waters. This result was expected since waters that were evaporated during convective drying (low-temperature drying) contain very little inorganic ionic compounds. All these values were significantly lower than the fixed norm for wastewater discharging in water bodies or reuse in agriculture. The blank tests (HS and HWC) showed a low impact of the sawdust and the wood chips since the related recovered waters had very low EC and COD compared to the same solid matrixes when impregnated with OMWW.

Concerning the COD contents, all the recovered water solutions had low values compared to the raw OMWW (100 g L^{-1}). The lowest value (2.1 g L^{-1}) was registered for impregnated sawdust (IS). The highest one (8.4 g L^{-1}) was observed for IWC and could be imputed to the release of supplementary organic matter from the solid matrix into the impregnating liquid phase. Electrical conductivity, pH, and COD tests were duplicated with recovered waters from different drying experiments. While pH and EC were close to each other, COD measurements were sensibly different (for example 2.3 and 14.4 g L^{-1} for OMWW). This difference is attributed to the time spent in open air. Indeed, the more the air was evacuated, the more the COD was decreased (2.3 g L^{-1} with 12 h venting and 14.4 g L^{-1} with no aeration). All the measured COD contents of the recovered waters were higher than the set standards. These values could be significantly decreased through specific treatment (depending on the detailed characterization of the contained dissolved organic matter) and mixing with other waters.

On the other hand, the IC analyses showed that for all the samples, the main anions and cations contents were significantly lower than the fixed norms (Table 2). Moreover, the suggested treatment cited above for the increase of these recovered waters pH by using low-cost alkaline materials such as powdered marble or seashell, will certainly enrich them with mineral elements such as calcium and magnesium.

Moreover, GC-MS investigations showed that the organic chemicals found came from olive oil, residues, and wood. Indeed, classical fatty acids coming from olive oil such as myristic, palmitic, and stearic acids, as well as tyrosol and vanillin were transferred into the recovered waters. Waters from OMWW and impregnated biomasses also contained short-chain acids such as acetic, malonic, butanoic, fumaric, and succinic, which could explain the difference of COD after sample venting.

Wood chips and sawdust significantly released chemicals like glycerol, beta-alanine, D-glucose, urea, and some short-chain organic acids (D-glyceric, succinic, acetic, etc.,) into waters. These chemicals come from both impregnated biomasses and humidified ones.

Therefore, a tertiary treatment is necessary to adjust the pH and decrease the COD in order to use these waters in agriculture. A low-cost solution could be the filtration through low-cost materials such as the cited-above raw/modified mineral wastes (seashell or powdered marble) and a proper ventilation to extract organic volatile compounds.

4. Conclusions

In this paper, a sustainable strategy for recovery of water from olive mill wastewater (OMWW) is proposed. This strategy includes (i) the impregnation of OMWW on lignocellulosic biomass (oak sawdust, wood chips), (ii) the controlled drying of the impregnation samples in a convective dryer, (iii) the water recovery from the drying operation through condensation process, and (iv) the material recovery of the dried impregnated biomasses for thermochemical conversion process (combustion/pyrolysis for energy/biochar production).

The quality of the recovered water was emphasized in this present work. In particular, different characterization techniques were used in order to assess the potential use of the recovered water in agriculture application. It was observed that the condensate waters are attractive for an agricultural purpose, but they do require an additional treatment to adjust the pH and to decrease its COD content.

This study performed at laboratory scale is a useful starting point for scaling up and ensuring sustainability.

Author Contributions: Conceptualization, P.D. and M.J.; methodology, P.D., M.-L.G. and S.J.; software, B.K.; validation, P.D., M.J. and S.J.; formal analysis, M.J. B.K. and M.-L.G.; investigation, P.D. and M.-L.G.; resources, P.D. and S.J.; data curation, P.D., M.J. and B.K.; writing—original draft preparation, B.K. and M.J.; writing—review and editing, P.D. and S.J.; visualization, P.D. and M.-L.G.; supervision, P.D., M.J. and S.J.; project administration, M.J.; funding acquisition, P.D., M.J. and S.J.

Funding: This work was partially funded by FERTICHAR project. FERTICHAR is funded through the ARIMNet2 (2017) Joint Call by the following funding agencies: ANR (France), HAO-DEMETER (Greece), MHESRT (Tunisia), ARIMNet2 (ERA-NET) has received funding from the European Union's Seventh Framework Programme for research, technological development and demonstration under grant agreement no. 618127. The authors gratefully acknowledge the funding agencies for their support. Authors would like also to thank the Region Alsace (now Grand Est) for its financial support concerning the drying unit pilot. This funding was given in the frame of the SEBASTE project (n° 131280).

Conflicts of Interest: The authors declare no conflict of interest.

References

1. Iglesias, A.; Garrote, L.; Flores, F.; Moneo, M. Challenges to Manage the Risk of Water Scarcity and Climate Change in the Mediterranean. *Water Resour. Manag.* **2007**, *21*, 775–788. [CrossRef]

2. Angelakis, A.N.; Marecos Do Monte, M.H.F.; Bontoux, L.; Asano, T. The status of wastewater reuse practice in the Mediterranean basin: Need for guidelines. *Water Res.* **1999**, *33*, 2201–2217. [CrossRef]

3. Mizyed, N.R. Challenges to treated wastewater reuse in arid and semi-arid areas. *Environ. Sci. Policy* **2013**, *25*, 186–195. [CrossRef]

4. Pedrero, F.; Kalavrouziotis, I.; Alarcón, J.J.; Koukoulakis, P.; Asano, T. Use of treated municipal wastewater in irrigated agriculture—Review of some practices in Spain and Greece. *Agric. Water Manag.* **2010**, *97*, 1233–1241. [CrossRef]

5. Khiari, B.; Wakkel, M.; Abdelmoumen, S.; Jeguirim, M. Dynamics and kinetics of cupric ion removal from wastewaters by Tunisian solid crude olive-oilwaste. *Materials* **2019**, *12*, 365. [CrossRef]

6. Wakkel, M.; Khiari, B.; Zagrouba, F. Textile wastewater treatment by agro-industrial waste: Equilibrium modelling, thermodynamics and mass transfer mechanisms of cationic dyes adsorption onto low-cost lignocellulosic adsorbent. *J. Taiwan Inst. Chem. Eng.* **2019**, *96*, 439–452. [CrossRef]

7. Khan, S.; Cao, Q.; Zheng, Y.M.; Huang, Y.Z.; Zhu, Y.G. Health risks of heavy metals in contaminated soils and food crops irrigated with wastewater in Beijing, China. *Environ. Pollut.* **2008**, *152*, 686–692. [CrossRef] [PubMed]

8. Wakkel, M.; Khiari, B. Basic red 2 and methyl violet adsorption by date pits: Adsorbent characterization, optimization by RSM and CCD, equilibrium and kinetic studies. *Environ. Sci. Pollut. Res.* **2018**, 1–19. [CrossRef] [PubMed]

9. Libutti, A.; Gatta, G.; Gagliardi, A.; Vergine, P.; Pollice, A.; Beneduce, L.; Disciglio, G.; Tarantino, E. Agro-industrial wastewater reuse for irrigation of a vegetable crop succession under Mediterranean conditions. *Agric. Water Manag.* **2018**, *196*, 1–14. [CrossRef]

10. Marrakchi, F.; Kriaa, K.; Hadrich, B.; Kechaou, N. Experimental investigation of processing parameters and effects of degumming, neutralization and bleaching on lampante virgin olive oil's quality. *Food Bioprod. Process.* **2015**, *94*, 124–135. [CrossRef]

11. Souilem, S.; El-Abbassi, A.; Kiai, H.; Hafidi, A.; Sayadi, S.; Galanakis, C.M. Chapter 1—Olive oil production sector: Environmental effects and sustainability challenges. In *Olive Mill Waste*; Galanakis, C.M., Ed.; Academic Press: Cambridge, MA, USA, 2017; pp. 1–28.

12. Negro, M.J.; Manzanares, P.; Ruiz, E.; Castro, E.; Ballesteros, M. Chapter 3—The biorefinery concept for the industrial valorization of residues from olive oil industry. In *Olive Mill Waste*; Galanakis, C.M., Ed.; Academic Press: Cambridge, MA, USA, 2017; pp. 57–78.

13. Kapellakis, I.E.; Tzanakakis, V.A.; Angelakis, A.N. Land application-based management of olive mill wastewater. *Water* **2015**, *7*, 362–376. [CrossRef]

14. Caputo, A.C.; Scacchia, F.; Pelagagge, P.M. Disposal of by-products in olive oil industry: Waste-to-energy solutions. *Appl. Therm. Eng.* **2003**, *23*, 197–214. [CrossRef]

15. Miranda, T.; Esteban, A.; Rojas, S.; Montero, I.; Ruiz, A. Combustion analysis of different olive residues. *Int. J. Mol. Sci.* **2008**, *9*, 512–525. [CrossRef] [PubMed]

16. Jeguirim, M.; Chouchène, A.; Réguillon, A.F.; Trouvé, G.; Le Buzit, G. A new valorisation strategy of olive mill wastewater: Impregnation on sawdust and combustion. *Resour. Conserv. Recycl.* **2012**, *59*, 4–8. [CrossRef]

17. Kraiem, N.; Jeguirim, M.; Limousy, L.; Lajili, M.; Dorge, S.; Michelin, L.; Said, R. Impregnation of olive mill wastewater on dry biomasses: Impact on chemical properties and combustion performances. *Energy* **2014**, *78*, 479–489. [CrossRef]

18. Haddad, K.; Jeguirim, M.; Jerbi, B.; Chouchene, A.; Dutournié, P.; Thevenin, N.; Ruidavets, L.; Jellali, S.; Limousy, L. Olive Mill Wastewater: From a Pollutant to Green Fuels, Agricultural Water Source and Biofertilizer. *ACS Sustain. Chem. Eng.* **2017**, *5*, 8988–8996. [CrossRef]

19. Jeguirim, M.; Dutournié, P.; Zorpas, A.A.; Limousy, L. Olive Mill Wastewater: From a Pollutant to Green Fuels, Agricultural Water Source and Bio-Fertilizer—Part 1. The Drying Kinetics. *Energies* **2017**, *10*, 1423.

20. Amos, W.A. *Report on Biomass Drying Technology*; Technical Report; National Renewable Energy Lab.: Golden, CO, USA, 1999. [CrossRef]

21. Borsato, A.V.; Doni-Filho, L.; Rakocevic, M.; Cocco, L.C.; Paglia, E.C. Chamomile essential oils extracted from flower heads and recovered water during drying process. *J. Food Process. Preserv.* **2009**, *33*, 500–512. [CrossRef]

22. Moore, W.A.; Kroner, R.C.; Ruchhoft, C.C. Dichromate Reflux Method for Determination of Oxygen Consumed. *Anal. Chem.* **1949**, *21*, 953–957. [CrossRef]

23. Montero, I.; Miranda, T.; Arranz, J.I.; Rojas, C.V. Thin Layer Drying Kinetics of By-Products from Olive Oil Processing. *Int. J. Mol. Sci.* **2011**, *12*, 7885–7897. [CrossRef]

24. Khiari, B.; Mihoubi, D.; Mabrouk, S.B.; Sassi, M. Experimental and numerical investigations on water behaviour in a solar tunnel drier. *Desalination* **2004**, *168*, 117–124. [CrossRef]

25. Selimi, A.; Nickelson, R. Effects of liming on soil pH and manganese toxicity in a Goulburn valley pear orchard. *Aust. J. Exp. Agric.* **1972**, *12*, 310–314. [CrossRef]

26. Haddad, K.; Jellali, S.; Jaouadi, S.; Benltifa, M.; Mlayah, A.; Hamzaoui, A.H. Raw and treated marble wastes reuse as low cost materials for phosphorus removal from aqueous solutions: Efficiencies and mechanisms. *C. R. Chim.* **2015**, *18*, 75–87. [CrossRef]

Article

Pyrolysis Process as a Sustainable Management Option of Poultry Manure: Characterization of the Derived Biochars and Assessment of their Nutrient Release Capacities

Samar Hadroug [1,2], Salah Jellali [3,*], James J. Leahy [4], Marzena Kwapinska [4], Mejdi Jeguirim [5], Helmi Hamdi [6] and Witold Kwapinski [4]

[1] Wastewaters and Environment Laboratory, Water Research and Technologies Center, P.O. Box 273, Soliman 8020, Tunisia; samarhadroug@gmail.com
[2] National Agricultural Institute of Tunisia, University of Carthage, Tunis 1082, Tunisia
[3] Center for Environmental Studies and Research, Sultan Qaboos University, P.O. Box 31, Al-Khoud 123, Muscat, Oman
[4] Department of Chemical Sciences, Bernal Institute, University of Limerick, V94 T9PX Limerick, Ireland; j.j.leahy@ul.ie (J.J.L.); marzena.kwapinska@ul.ie (M.K.); witold.kwapinski@ul.ie (W.K.)
[5] CNRS, IS2M UMR 7361, University of Haute-Alsace, University of Strasbourg, F-68100 Mulhouse, France; mejdi.jeguirim@uha.fr
[6] Center for Sustainable Development, College of Arts and Sciences, Qatar University, P.O. Box 2713 Doha, Qatar; hhamdi@qu.edu.qa
* Correspondence: s.jellali@squ.edu.om; Tel.: +968-2414-1807

Received: 10 September 2019; Accepted: 28 October 2019; Published: 30 October 2019

Abstract: Raw poultry manure (RPM) and its derived biochars at temperatures of 400 (B400) and 600 °C (B600) were physico-chemically characterized, and their ability to release nutrients was assessed under static conditions. The experimental results showed that RPM pyrolysis operation significantly affects its morphology, surface charges, and area, as well as its functional groups contents, which in turn influences its nutrient release ability. The batch experiments indicated that nutrient release from the RPM as well as biochars attains a pseudo-equilibrium state after a contact time of about 48 h. RPM pyrolysis increased phosphorus stability in residual biochars and, in contrast, transformed potassium to a more leachable form. For instance, at this contact time, P- and K-released amounts passed from 5.1 and 25.6 mg g^{-1} for RPM to only 3.8 and more than 43.3 mg g^{-1} for B400, respectively. On the other hand, six successive leaching batch experiments with a duration of 48 h each showed that P and K release from the produced biochars was a very slow process since negligible amounts continued to be released even after a total duration of 12 days. All these results suggest that RPM-derived biochars have specific physico-chemical characteristics allowing them to be used in agriculture as low-cost and slow-release fertilizers.

Keywords: raw poultry manure; pyrolysis; biochar; characterization; leaching; phosphorus; potassium

---

1. Introduction

Raw poultry manure (RPM) is a chicken waste that is produced from the normal operation of hatcheries, turkey, broiler, and egg laying [1]. Huge amounts of RPM are annually produced in the world. In the USA, the annual production of animal manure dry matter was estimated to be more than 19 million tons in 2018 [2]. In Tunisia, poultry production has seen a sharp increase during the last decade, generating about 400,000 tons/year of manure [3].

The composition of poultry manure is variable and depends mainly on the type of feed [4]. It generally contains high levels of organic carbon and nutrients, such as nitrogen, potassium, and phosphorus, which are indispensable macro-elements for soil fertility improvement and crop growth enhancement [5,6]. The sustainable management of this biowaste has been pointed out as an urgent priority due to its possible negative effects on human health and the environment [5,7]. In fact, the traditional methods for RPM management have various drawbacks [8]. Incineration has been the most widely used technique to reduce the huge volumes of RPM. However, the relatively high ash contents in RPM requires mitigation technologies [9]. Thermal gasification has also been investigated at the farm level, but gas quality is poor and requires extensive cleaning, which is expensive, particularly at the farm scale [10]. Likewise, the methane-fermentation process of malodorous RPM is generally limited by its excessive nitrogen contents [3]. Direct land application of fresh RPM or its composted form as organic amendments may result in: (i) The spread of pathogens [1], (ii) air pollution due to the emission of greenhouse gases and phytotoxic substances [11,12], and (iii) possible surface water resources eutrophication and groundwater pollution [3] due to nitrogen and phosphorus leaching [13]. One of the main potential methods used to reduce nutrient losses from RPM is thermal conversion by pyrolysis (in the absence of oxygen) into biogas and bio-oil that could have an important energetic added value [14–16] and a solid residue, named biochar. Biochars are highly porous carbonaceous materials that could be valorized for agricultural purposes as soil amendments [17,18], for climate change mitigation through carbon sequestration and a reduction of greenhouse gas emissions [19], and for environmental protection since they exhibit large capacities in removing various pollutants from both liquid and gas effluents [2,14,20,21]. All the above-cited effects are very dependent on the physico-chemical characteristics of the biochars that are mainly linked to the nature of the used feedstock and also to the pyrolysis experimental settings [22,23]. Therefore, in-depth characterization of produced biochars under given conditions has been pointed out as a key step in order to select the best method for their valorization [24]. On the other hand, according to our best of knowledge, no studies have investigated controlled RPM-derived biochars' capacities in releasing nutrients versus time by using the same biochars in successive leaching assays, which is close to what is happening in situ where biochar-amended soils are successively leached with rainfall and irrigation events. The majority of studies have attempted to monitor nutrient-release kinetics in a continuous way [22,24,25]. On the other hand, data on the relationships between the physico-chemical characteristics of RPM and RPM-derived biochars and their time-dependent capacity for phosphorus (P) and potassium (K) release in aqueous solutions are still missing.

Accordingly, the aims of the current research work were: (i) To study the thermal properties of RPM through thermogravimetric analyses, (ii) to determine the role of the selected pyrolysis temperatures on the physico-chemical properties of produced biochars, and (iii) to assess the importance of physico-chemical characteristics of RPM and its derived biochars on the kinetics and efficiency of phosphorus and potassium release under various experimental conditions.

2. Materials and Methods

2.1. Preparation of Raw Poultry Manure

The raw poultry manure was collected from a poultry farm in the city of Mornag (South of Tunis, Tunisia). Fresh RPM feedstock was first air-dried for about 10 days until a constant weight. Then, it was manually ground in order to have a relatively homogeneous particle size. Finally, RPM was mechanically sieved by using mechanical sieve shaker (Retch, Haan, Germany). Only fraction with a particle size less than 1 mm was collected and stored in bags for subsequent analysis.

2.2. Thermogravimetric Analysis of RPM

Thermogravimetric analysis (TGA) of RPM was performed using a thermal analyzer (LABSYS TG brand, Ireland) under the following thermo-analytical conditions: An RPM mass of 16.7 mg, a heating gradient of 10 °C min^{-1} for temperatures varying from 30 to 810 °C, and a dynamic nitrogen atmosphere with a flow rate of 10 mL min^{-1}. These TGA measurements were carried out in triplicate and the mean values are reported in this study. The TG curves giving the variation of the residual RPM mass (%) versus time/temperature were derived (DTG curves) in order to get a better idea about the kinetic of RPM mass losses (mg min^{-1}) and to highlight even slight mass changes.

2.3. Biochar Production and Characterization

The biochars used in this study were derived from pyrolysis of air-dried RPM at two different temperatures (400 and 600 °C). These biochars (named B400 and B600, consecutively) were produced using a laboratory pyrolysis furnace (Lenton brand oven) with dimensions of 67 cm length, 23 cm width, and 22 cm height. A pyrolysis metal container with a 13 cm length and 11 cm width equipped with four 0.5-cm holes in the lid was used, which permitted vapor elimination and maintained an air-absent environment in the container during the pyrolysis process. At the end of the pyrolysis operation, the container was taken out of the furnace and cooled down to room temperature. Solid residues (biochars) were removed from the container and placed in a plastic bag for use throughout this study.

Air-dried RPM, B400, and B600 were physically characterized through an assessment of: Particle size distribution by using a Malvern Mastersizer STD06 laser granulometry; surface area and pore volumes according to the Brunauer, Emmet, and Teller (BET) method, via the measurements obtained by N$_2$ adsorption using an Autosorb Quantachrome apparatus; and finally, their morphology and structure through a scanning electron microscope (SEM) apparatus (SU-70). These solid matrices were also chemically characterized by the determination of: (i) Crystalline phase through X-ray diffraction (XRD) analyses; (ii) pH of zero-point charge (pHZC) according to the methodology given by Azzaz et al. [26]; (iii) elemental composition (C, H, N, O, S) by using a "vario EL cube" elemental analyzer; (iv) moisture, volatile matter, ash, and fixed carbon contents according to the British Standard Institution (BSI) protocol; and (v) their main contained functional groups through a Fourier transform infrared spectroscopy apparatus (Agilent Tech Cary 630, Ireland). The spectral resolution of the FTIR was 1 cm^{-1} measured between 400 and 4000 cm^{-1}.

2.4. Batch Nutrient Release Experiments

The batch experiments consisted of determining the P and K release kinetics from RPM and its derived biochars, as well as the effect of some key parameters, such as solution pH and solid matrices dosages, on this release process. These experiments were conducted at 20 ± 2 °C in 120-mL capped flasks. During these assays, a defined amount of RPM or biochars was shaken in 50 mL of distilled water for a fixed time at 400 rpm using an IKA RT15 Power IKAMAG multi position magnetic stirrer for a contact time 48 h. Then, the resulting suspensions were centrifuged at 3000 rpm for 8 min using a ROTOFIX 32 apparatus and filtrated using a vacuum pumping system. Dissolved P and K concentrations in filtrates were analyzed by an inductively coupled plasma spectrometry (ICP) device (Agilent Tech 5100 ICP OES, Ireland). The released amount of P "q_P (mg g^{-1})" or K "q_K (mg g^{-1})" per gram of solid matrix were determined as follows:

$$q_P = \frac{C_P}{D}, \tag{1}$$

$$q_K = \frac{C_K}{D}, \tag{2}$$

where C_P, C_K, *and* D are the measured released concentrations of P and K, and the used solid matrix dose, respectively.

All the following batch nutrient release assays were carried out in triplicate and each analysis data reported in this study is an average of at least three independent parallel sample solutions.

2.4.1. Effect of Contact Time

The kinetics of P and K release from RPM and the two derived biochars (B400 and B600) was monitored at 1, 5, 10, and 30 min and 4, 17, 24, and 48 h. During these assays, the solid matrices dosage, initial aqueous pH, and temperature were fixed to 10 g L^{-1}, 5.6 (natural without adjustment), and $20 \pm 2\,°C$, respectively. The subsequent nutrient relse data were fitted to the most common models, namely pseudo-first order (PFO) and pseudo-second order (PSO) models. The corresponding equations and the related assumptions were given by Azzaz et al. [26]. In order to assess the applicability of these two models and their ability in fitting the current experimental data, average percentages errors (APEs) between the measured and calculated released amounts were estimated as follows [27]:

$$\text{APE} = \frac{\sum_{i=1}^{N} \left| \frac{q_{exp,i} - q_{th,i}}{q_{exp,i}} \right|}{N}, \tag{3}$$

where $q_{exp,i}$ and $q_{th,i}$ are the experimental and theoretical released P or K quantities at a given time "i", respectively. N is the number of experimental data.

2.4.2. Effect of Initial Aqueous pH

The effect of initial aqueous pH on the rate of P and K release from RPM and the derived biochars was assessed for adjusted values of 3, 5.6 (natural pH), and 8. All these experiments were carried out at a constant dose of 10 g L^{-1}, a temperature of $20 \pm 2\,°C$, and a contact time of 48 h. The aqueous pH adjustment was performed by using 0.1 M HNO$_3$ or NaOH solutions.

2.4.3. Effect of RPM and RPM-Derived Biochars Doses

The impact of RPM, B400, and B600 doses on P and K release was assessed for an initial aqueous pH of 5.6 (without adjustment), a temperature of $20 \pm 2\,°C$ and a contact time of 48 h. The tested doses were 2, 5, 10, and 15 g L^{-1}.

2.4.4. Successive Nutrient Release Experiments

In order to determine the slowness of the RPM and its two derived biochars' capacities in releasing P and K, six successive batch experiments were carried out for a total contact time duration of 12 days. For each leaching experiment, the same solid samples were put in contact with distilled water at a dose of 10 g L^{-1}, an initial pH of 5.6 (without adjustment), and a contact time of 48 h. At the end of each assay, these solid matrices were recovered through filtration by using a vacuum pumping system with 0.45-μm paper filters, and then dried overnight at $40\,°C$. These dried samples were reused for the next leaching experiment in the presence of new distilled water under the same experimental conditions cited above.

2.5. Statistical Analysis

All the experimental data (except elemental analysis and granulometric distribution) for RPM and its derived biochars represent means of at least triplicate measurements. Data were analyzed using STATISTICA 8.0 (Statsoft, Tulsa, OK, USA). ANOVA with Duncan's multiple range test were applied for mean separation at $p < 0.05$.

3. Results and Discussion

3.1. Thermogravimetric Analysis of Raw Poultry Manure

The TG analysis of RPM was carried out under the experimental conditions given in Section 2.2. Figure 1 reports the RPM's TG records and the corresponding first derivative curve (DTG).

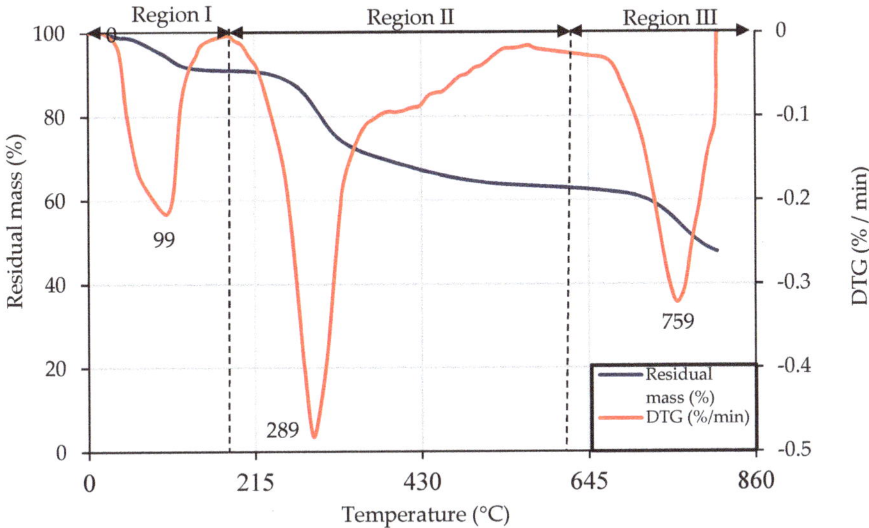

Figure 1. Thermograms (TGs) and its first derivative (DTG) of the used raw poultry manure (RPM dose = 16.7 mg; heating gradient = 10 °C min^{-1}).

Figure 1 shows the presence of three regions with different thermal behavior. Region I spans a temperature interval of 49 to 179 °C, with a progressive mass loss reaching a maximum rate of 9% at the end of this interval. In this region, the maximum loss rate (0.22% min^{-1}) was observed at a temperature of 99 °C and corresponds to moisture losses from the RPM. The second region is comprised between 179 and 609 °C, with a corresponding mass loss of about 27%. A maximum peak rate (0.48% min^{1}) was observed at a temperature of 289 °C. This peak apparition is mainly due to the thermo-oxidative degradation of highly oxygenated organic matter contained in RPM, such as labile alkyl systems, carbohydrates, and fatty acids [28,29]. The last thermal region corresponds to temperatures ranging between 609 to 810 °C, with a corresponding mass loss of 15%. A peak degradation rate was observed at a temperature of 759 °C (0.32% min^{-1}) and could be attributed to the oxidation of mineral and biogenic salts, such as calcium and potassium carbonate [30]. A global similar trend was reported by Cimo et al. [30] when studying the thermal behavior of RPM collected from a Sicilian chicken farm in Italy. It is important to underline that P speciation in biochars is very dependent on the pyrolysis temperature. At low temperatures, the organic carbon compounds are aromatic and therefore do not overlap with the P vibrations. For biochars produced at higher temperatures, prevailing P forms are tricalcium phosphate, hydroxyapatite, calcium, and iron phosphate precipitates [31].

3.2. Characterization of Raw Poultry Manure and Biochars

The main physico-chemical properties of RPM as well as its two derived biochars at temperatures of 400 and 600 °C are given in Table 1. Granulometric characterization showed that the RPM and the corresponding biochars have a relatively fine texture with a mean particle size of 0.48, 0.43, and 0.31 mm, respectively. The uniformity coefficient (UC), which is the ratio of d60/d10, showed that RPM and B400 are heterogeneous media because their UCs are higher than 2. However, B600 could be considered as homogenous since its UC is lower than 2. The fine texture of the relative RPM and derived biochars indicates that they could play an important role in physico-chemical reactions with pollutants contained in rainwater or irrigation water [32].

Table 1. Main physico-chemical characteristics of raw poultry manure (RPM), B400, and B600 (*: In Wt.% dry basis). For each parameter, means with the same lowercase letters are not statistically different at $p < 0.05$.

Physico-Chemical Property	RPM	B400	B600
Grains Size Distribution			
d_{10} (mm)	0.21	0.16	0.19
d_{50} (mm)	0.48	0.43	0.31
d_{60} (mm)	0.57	0.44	0.33
Uniformity coefficient (UC = d_{60}/d_{10})	2.71	2.75	1.74
Surface area ($m^2\ g^{-1}$)			
BET	0.88	4.30	5.34
Charges density			
pHZC	9.09 ± 0.03 a	10.87 ± 0.008 b	11.47 ± 0.015 c
Ultimate Analysis (%)*			
C	25.56 ± 1.39 b	22.04 ± 1.75 ab	21.22 2.30a
H	3.27 ± 0.79 b	0.87 ± 0.015 a	0.54 ± 0.04 a
N	2.19 ± 0.17 b	0.95 ± 0.19 a	0.64 ± 0.11 a
S	0.69 ± 0.0.17 a	0.68 ± 0.46 a	0.66 ± 0.005 a
O	69.35± 2.43 a	75.45 ± 2.31 b	76.90 ± 2.38 b
C/H	6.87 ± 1.85 a	25.33 ± 1.59 b	35.53 ± 7.61 c
Mineral Composition ($mg\ g^{-1}$)			
Al	15.97	18.30	23.22
Fe	2.24	3.46	3.68
K	38.12	71.95	66.23
Mg	4.62	6.64	7.11
Na	18.98	20.17	28.67
P	17.00	20.33	43.17
Si	265.81	277.85	390.95
Ti	0.37	0.39	0.51
Proximate Analysis (%)*			
Moisture	16.69 ± 0.04 b	0.76 ± 0.09 a	0.80 ± 0.04 a
Ash	51.35 ± 0.38 a	78.64 ± 0.62 b	79.74 ± 0.18 c
Volatile matter	36.34 ± 0.18 b	4.08 ± 0.12 a	3.95 ± 0.13 a
Fixed carbon	11.99 ± 0.05 a	17.72 ± 0.13 c	16.18 ± 0.24 b

The BET surface area of RPM was assessed as 0.88 $m^2\ g^{-1}$. Pyrolysis operation increased this value to 4.30 and 5.34 $m^2\ g^{-1}$ for B400 and B600, respectively (Table 1). These relatively low surface areas could be imputed to the existence of inorganic ash materials at high concentrations that might fill or block access to micropores [22,33]. Generally, higher specific surfaces areas are obtained for higher carbonization temperatures of biomasses [23,24]. It is worth mentioning that the BET surface areas of biochars significantly affect not only the subsurface contaminant mobility but also the microbial activities [22,29]. These authors reported that the BET specific surface areas of their RPM and derived biochar at 400 °C were 0.90 and 3.94 $m^2\ g^{-1}$, respectively, which are fairly close to the values found in the current study.

SEM analyses of the RPM and its derived biochars produced at temperatures of 400 and 600 °C are shown in Figure S1. RPM particles present a rough structure with a relatively oval shape (Figure S1). Besides, the pyrolysis process did not induce significant morphological changes. The presence of relatively small-sized particles in the three used solid materials (Figure S1) should play an important role in the nature of the physicochemical reactions with dissolved substances in liquid effluents [34].

The XRD spectra of RMP, B400, and B600 are given in Figure S2. Raw poultry manure showed an elevated background between 2θ = 21°, likely attributable to the organic matter content [35,36]. The peaks at 2θ = 28.9° and 2θ = 29.6° indicate the presence of low crystallinity (quartz and calcite) that already exists with a low intensity compared to its two biochars produced at 400 and 600 °C, respectively [37]. In addition to quartz and calcite, the sylvite (KCl), which is a cubic mesh mineral species of the chloride family, is present for B400 and B600 at peaks equal to $2\theta = 40.5°$. The phosphate mineral whitlockite ($(Ca,Mg)_3(PO_4)_2$) was identified only in B600 at a peak equal to $2\theta = 47.3°$ since phosphorus reached a more stable form [24,35].

The measured pHZC values of RPM, B400, and B600 were 9.09, 10.87, and 11.47, respectively (Table 1). This significant increase is mainly attributed to the concentration of the non-carbonized inorganic elements that were already present in the original feedstock [38]. The same behavior was reported by Song and Guo [22], who found that RPM carbonization at temperatures of 400 and 600 °C increased pHZC from 9.5 to 10.1 and 11.5, respectively. This observed alkaline property implies that these materials could be used as pH regulators for acidic soils when used as amendments or for acidic wastewaters when used as adsorbents for toxic substances. Such valorization pathways will significantly reduce the use of chemical reagents and hence the related expenses.

On the other hand, RPM pyrolysis significantly decreased C, H, N, and S contents as temperature increased (Table 1). For instance, original C, H, N, and S contents in RPM decreased respectively by 5.7%, 3.2%, 1.6%, and 0.02% in B600. These findings are mainly due to their transfer to bio-oil and/or incondensable exhaust during the pyrolysis process [24]. Furthermore, the C/H ratio increased with the increase of the pyrolysis temperature (Table 1) as a result of dehydration and decarboxylation reactions [33]. A similar trend was reported by Cimo et al. [30] when studying the pyrolysis of RPM collected from a chicken farm. They found C/H ratios of 7, 21, and 42 for RPM, and derived biochars at temperatures of 450 and 600 °C, respectively. The larger the C/H ratio is, the higher the polycondensation degree of organic systems. This confirms that an increasing pyrolysis temperature favors the formation of polycondensation aromatic rings.

The analysis of inorganic elements showed that silicon (Si), potassium (K), sodium (Na), phosphorus (P), aluminum (Al), and magnesium (Mg) exist in RPM with relatively high contents. The concentrations of these elements increase when increasing the pyrolysis temperature (Table 1). For instance, initial Mg, P, and K contents in RPM increased by 43.7% and 53.9%, 19.5% and 153.9%, 88.7% and 73.7% for B400 and B600, respectively. The presence of these nutrients in the produced biochars would be very useful for plant development and growth after land application.

Volatile matter (VM) contents significantly decreased from 36.3% for RPM to only 4.1% and 4.0% for B400 and B600. These changes are mainly due to the conversion of the RPM's organic matter into gaseous products during the pyrolysis process [39]. In contrast, ash contents increased from 51.4% for RPM to 78.6% and 79.7% for B400 and B600, respectively. The high ash contents of the three materials might be attributed to the physically incorporated soils and intrinsic inorganic elements, such as phosphorus and potassium, that are present in the raw manure and derived biochars [1]. Furthermore, the increase of the pyrolysis temperature leads to the dissociation of minerals from biomass [40]. As for ash, the fixed carbon contents increased after RPM pyrolysis, resulting in a more stable phase [39,41]. Unexpectedly, the measured fixed carbon content at 600 °C was slightly higher than the one measured at 400 °C (5.6%), which could be due to a combination of the samples' heterogeneity and analyses' uncertainties. Similar findings were reported by Enders et al. [42]. Finally, due to water evaporation during the pyrolysis process, the measured RPM moisture decreased from 16.7% to about 0.8% for both B400 and B600. Similar trends were also reported by Akdeniz and Novais et al. [2,37].

The FTIR spectra of raw manure feedstock and both biochars are given in Figure S3. For raw poultry manure, the major band near 3853 cm^{-1} (H-bonded OH) was attributed to the stretching vibration of hydrogen bonds of carboxylic acid and alcohols [30]. Peaks at 2992 cm^{-1} and 2865 cm^{-1} revealed asymmetric and symmetric (–CH) stretching bands of CH_3 and CH_2 groups, respectively and are associated with the aliphatic functional groups existing in lipids [43]. The peak at 1650 cm^{-1} concerns the C=O stretching of CO_2H and C=O of primary amides, which indicates the presence of protein. A CH_2 deformation peak registered at 1416 cm^{-1} is characteristic of saturated fatty acids and cellulose [44]. The peak at 1233 cm^{-1} of C-O is associated with the stretching of carbohydrates. Finally, the small peak appearing between 871 and 1024 cm^{-1} in the raw feedstock was attributed to stretching -CH and represents the aromatic structures' out-of-plane bending vibrations [39].

The pyrolysis of RPM at 400 and 600 °C generated new transformational products. Compared to RPM, the spectra of produced biochars reveal the presence of higher aromatic and lower carbohydrates' structures. In this context, the intensity of the H-bonded OH peak significantly decreased with the increase of the pyrolysis temperature. This is likely related to the dehydration process during the pyrolysis operation and suggests that the carboxylic acid might be dissolved and hydrolyzed. Furthermore, the aliphatic C-H stretching of CH$_3$ and CH$_2$ groups observed respectively at 2992 and 2865 cm^{-1} almost disappeared in the biochars' spectra, which could be imputed to the hydrolysis of aliphatic chain and dissolution. Similarly, the peak at 1650 cm^{-1}, which is associated with C=O stretching of primary amides, presents a clear intensity decrease with an increasing pyrolysis temperature. This behavior is probably due to the decomposition of the C=O bonding that generally enhances the formation of monoxide carbon and dioxide carbon [43], and causes a change in N functional groups by pyrolysis [45,46]. The observed peak at 1416 cm^{-1}, which is associated with CH$_2$ deformation for both biochars, presents an intensity increase with temperature [39]. Also, the bonding of C-O increased at 1233 cm^{-1}, which is associated with the stretching of –COH in alcohols and –COR in aliphatic ethers [44]. The peak between 871 and 1024 cm^{-1} in the raw poultry manure is associated to aromatic –CH out-of-plane bending vibrations. This peak increased with the pyrolysis temperature, indicating a gradual conversion of aromatic structures during this process [44]. The presence of the various above-cited functional groups may play an important role in nutrient leaching from these biochars and especially pollutant adsorption from wastewaters [45,46]. For instance, P ion adsorption was confirmed by the apparition of a new P-O stretching vibration at 1050 cm^{-1} [47].

3.3. Phosphorus and Potassium Release

The release of P and K from RPM and biochars was assessed in the batch mode under the various experimental conditions detailed in Section 2.4.

3.3.1. Effect of Contact Time—Kinetic Release Study

In order to assess the effect of contact time on the release efficiency of P and K from RPM, B400, and B600, batch experiments were performed for an initial pH of 5.6 (without adjustment), a constant dose of 10 g L^{-1}, and contact times varying between 1 min and 48 h (see Section 2.4.1). Outcomes indicated that P and K release from the three studied solid matrices was clearly a time-dependent process (Figure 2a,b and Figure 6a,b).

The P release kinetics were fast at the beginning of the experiment (until 4 h) as shown in Figure 2b. This is probably attributed to an ion exchange or desorption process and/or dissolution of some poorly crystalline or amorphous phosphates [25]. At this time, the released amount was estimated to 4.81 mg g^{-1}, which represents 94% of the overall released quantity. Then, the released amount continued to increase but at much slower rates until an equilibrium state was reached. This latter state, which is characterized by an approximate released amount, was observed after approximately 17 h. A similar trend was also observed for potassium release (Figure 3a,b). The observed stiff slope at the beginning of the experiments indicates a quick and instantaneous release related to the abundance of low-energy-retained P or K (P/K) elements on the surface of the solid matrices. At this step, the release rate was relatively fast because P/K was essentially desorbed from the easily accessible sites to water located at the exterior surfaces of RPM and the derived biochars. When the release of P/K from the exterior surfaces reached its limits, P/K could be further released from the more inaccessible sites located at the inner surfaces of the tested materials.

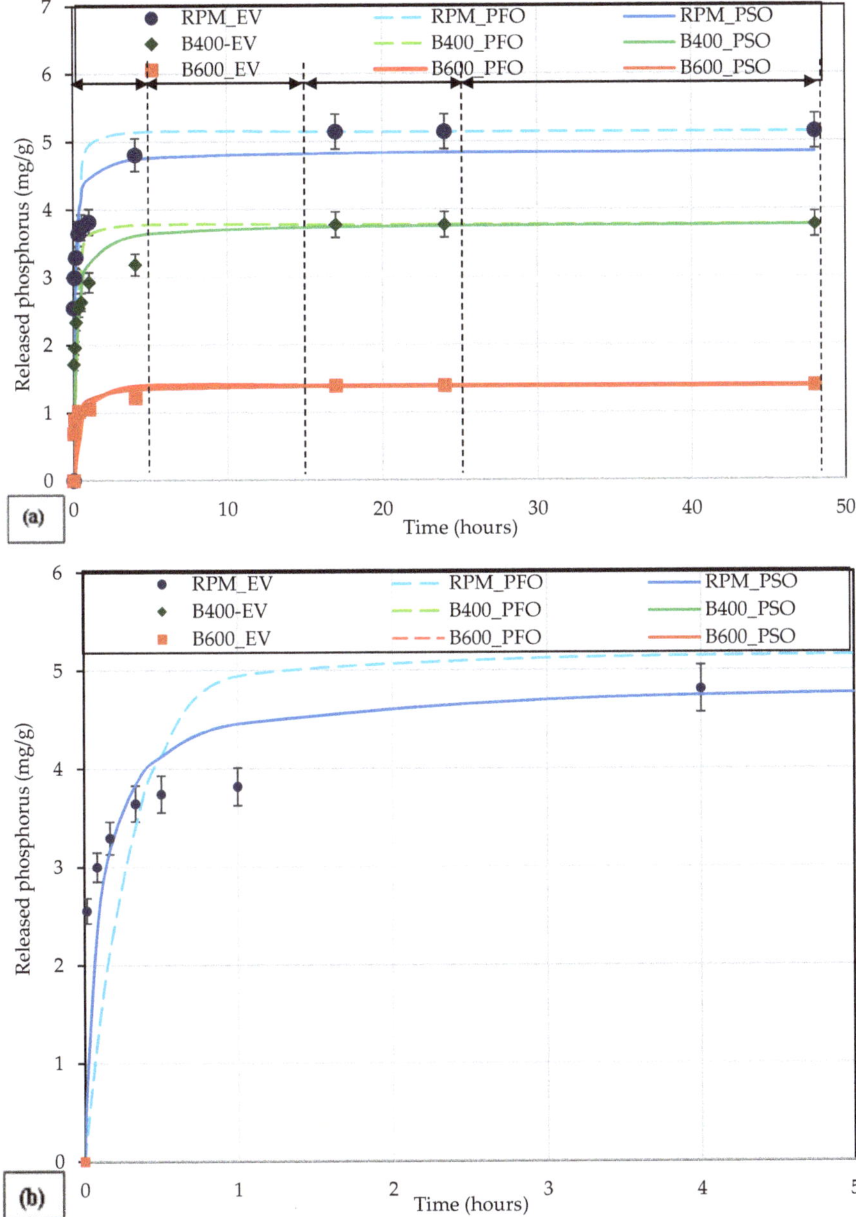

Figure 2. Comparison between the experimental quantities and theoretical ones of phosphorus released by raw poultry manure (RPM) and its derived biochars (EV: Experimental values; PFO: pseudo-first order; PSO: pseudo-second order) for 48 h (**a**) and zoom for 5 h (**b**).

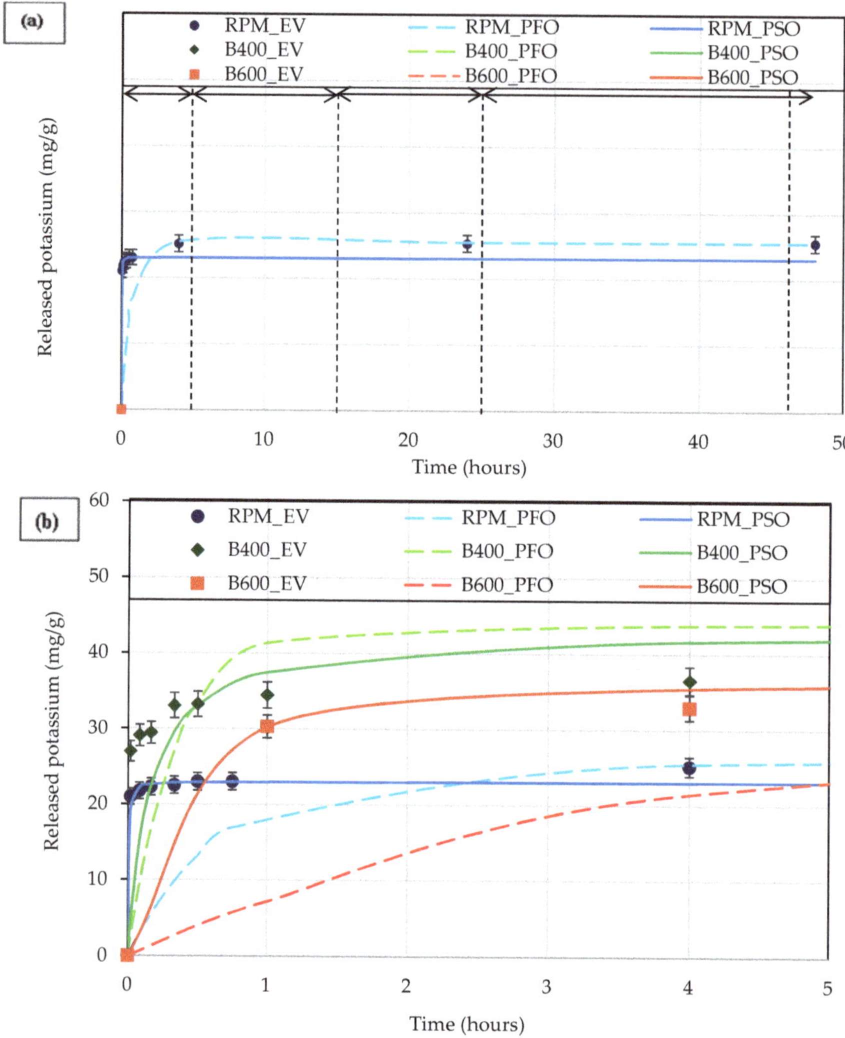

Figure 3. Comparison between the experimental and theoretical quantities of potassium released by raw poultry manure (RPM) and its derived biochars (EV: Experimental values; PFO: pseudo-first order; PSO: pseudo-second order) for 48 h (**a**) and zoom for 5 h (**b**).

Even if the pyrolysis process had concentrated P in the biochars compared to RPM (Table 1), at equilibrium, the released P amounts from RPM (5.14 mg g^{-1}) were much higher than those corresponding to B400 (3.77 mg g^{-1}) and B600 (1.39 mg g^{-1}). These amounts correspond to leachable P percentages of 30.2%, 18.5%, and 3.2% for RPM, B400, and B600, respectively. These rates are likely due to the fact that the pyrolysis process had converted a portion of RPM-containing phosphorus into water non-extractable forms, such as crystalline metal phosphates, which generally dissolve much more slowly and are less available [45,48]. In this context, Wang et al. [24] reported that RPM wastes collected from local hen houses (USA) released an average phosphorus amount of 6.48 mg g^{-1}. This quantity decreased to about 1.04 mg g^{-1} for its derived biochar at a temperature of 600 °C. Similarly, Peak et al. [48] also reported that the released amount of phosphorus from an RPM sample (collected from the region of Avila, Spain) was about 2.57 times higher than the one measured for its derived biochar produced at 500 °C.

The required time to attain an equilibrium state for K was much longer compared to P since it took about 48 h of contact time (Figure 3a). Furthermore, it appears that the pyrolysis process converted the initial potassium in RPM into a more leachable and water-soluble form in biochars through dissociation mechanisms [40]. As such, released K amounts were 25.6, 43.3, and 37.7 mg g^{-1} for RPM, B400, and B600, respectively. Accordingly, Song and Guo [22] showed that the released potassium amount from an RPM-derived biochar at a temperature of 600 °C (44.61 mg g^{-1}) was 1.9 higher than that observed for the raw biomass.

According to Tables 2 and 3, and Figures 2b and 3b, the confrontation of P/K release experimental data to those determined by the pseudo-first-order (PFO) and pseudo-second-order (PSO) models showed that the PFO model completely fails to reproduce the experimental values for the three tested solid materials. In fact, the calculated P/K determination coefficients as well as the related average percentage errors between the measured and the calculated values were not at all satisfactory. For instance, the "determination coefficients" and the APE of P were two by two 0.786 and 23.7%, 0.822 and 23.4%, and 0.734 and 29.6% for RPM, B400, and B600, respectively.

Table 2. Kinetic parameters of phosphorus release by RPM and its two derived biochars at 400 and 600 °C.

Solid Support	Pseudo-First-Order Model (PFO)			Pseudo-Second-Order Model (PSO)		
	$K_{1.PFO}$ (h^{-1})	R^2_{PFO}	APE$_{PFO}$ (%)	$K_{2.PSO}$ (g mg^{-1} h^{-1})	R^2_{PSO}	APE$_{PSO}$ (%)
RPM	3.238	0.786	23.714	2.318	0.867	14.467
B400	3.242	0.822	23.431	1.435	0.879	17.766
B600	1.688	0.734	29.566	4.602	0.836	19.860

Table 3. Kinetic parameters of potassium release by RPM and its two derived biochars at 400 and 600 °C.

Solid Support	Pseudo-First-Order Model (PFO)			Pseudo-Second-Order Model (PSO)		
	$K_{1.PFO}$ (h^{-1})	R^2_{PFO}	APE$_{PFO}$ (%)	$K_{2.PSO}$ (g mg^{-1} h^{-1})	R^2_{PSO}	APE$_{PSO}$ (%)
RPM	3.618	0.474	27.130	1.087	0.743	17.837
B400	2.886	0.668	26.199	0.150	0.750	19.517
B600	0.039	0.422	56.748	0.109	0.995	1.737

On the contrary, the PSO model succeeded to sufficiently reproduce the experimental data since the calculated determination coefficients (R2) for P or K were higher than that of the pseudo first order. Moreover, the calculated APE between the experimental and theoretical released P or K values were significantly lower than those calculated by the PFO model (Tables 2 and 3). For the three solid materials and the PSO model, the maximum values of APE for P and K were estimated at 19.9% (Table 2) and 19.5% (Table 3), respectively. However, the maximum values of this parameter were 29.6% and 56.7% for P and K, respectively. The highest non-accordance between the experimental and theoretical data corresponds to the initial phases of the experiments where the nutrient release slopes are maximal (Figures 2b and 3b). The fact that the experimental data were better fitted by the PSO suggests that nutrient leaching from RPM and its derived biochars might be controlled by a chemical process [26].

3.3.2. Effect of pH

The effect of initial aqueous pH on P and K release from the three materials was assessed under the experimental conditions described in Section 2.4.2. The results are illustrated in Figure 4a,b.

P-released amounts from RPM were approximately constant, with an average value of 5.51 mg g^{-1}. However, for B400 and B600, P-released amounts increased with the increase of pH values. For instance, the released amount from B600 increased by about 43.5% when the pH increased from 3 to 8. Globally, a similar variation trend was observed for K. The maximum released amount (41.33 mg g^{-1}) was observed for B400 at an initial aqueous leaching pH of 8. It is worth mentioning that high aqueous pH values could lead to the precipitation of the released phosphorus and potassium and therefore affect their availability to plants. For instance, depending on the aqueous pH value, phosphorus can exist as tricalcium phosphate, hydroxyapatite, calcium, and iron phosphates compounds [31]. However, it is very difficult for the precipitation of potassium as potassium hydroxide, which could be observed for very high pH values, to occur under the tested experimental conditions. It is important to underline that for high pH values, phosphorus could precipitate into calcium phosphate and magnesium phosphate, which makes P assimilation by crops more difficult [34,49]. These results confirmed that for a wide pH range, the biochars derived from RPM represent an important source of potassium that could be assimilated by various crops after soil application.

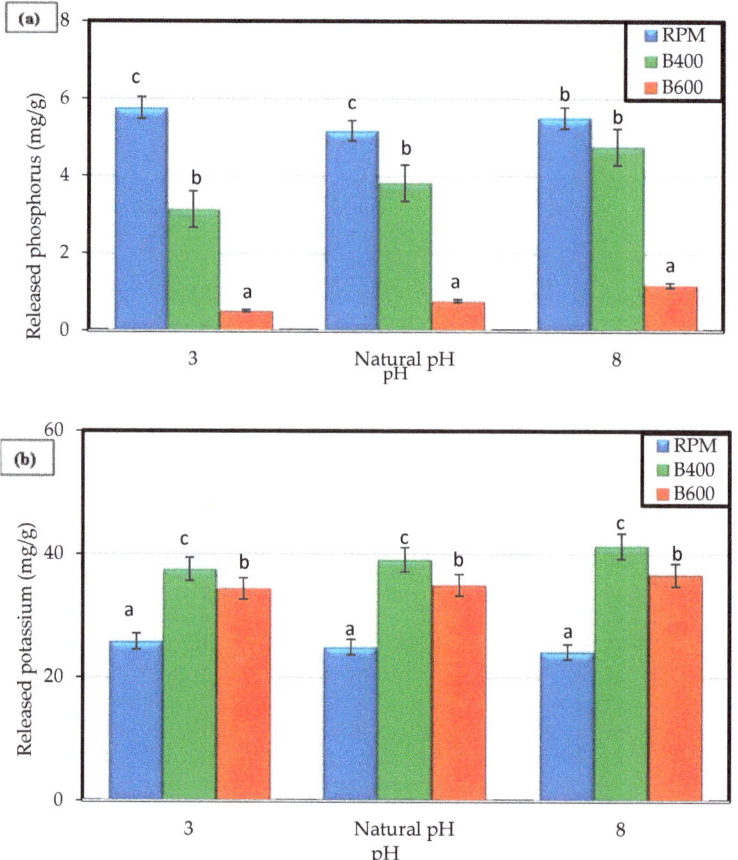

Figure 4. Effect of initial aqueous pH on the phosphorus (**a**) and potassium (**b**) release efficiency from raw poultry manure and its biochars derived at temperatures of 400 and 600 °C (dosage = 10 g L^{-1}, T = 20 ± 2 °C). At each pH value, means with the same lowercase letters are not statistically different at $p < 0.05$.

3.3.3. Effect of Dose

The effect of RPM, B400, and B600 doses on P and K release efficiency was determined according to the experimental conditions given in Section 2.4.3 and are illustrated in Figure 5a,b.

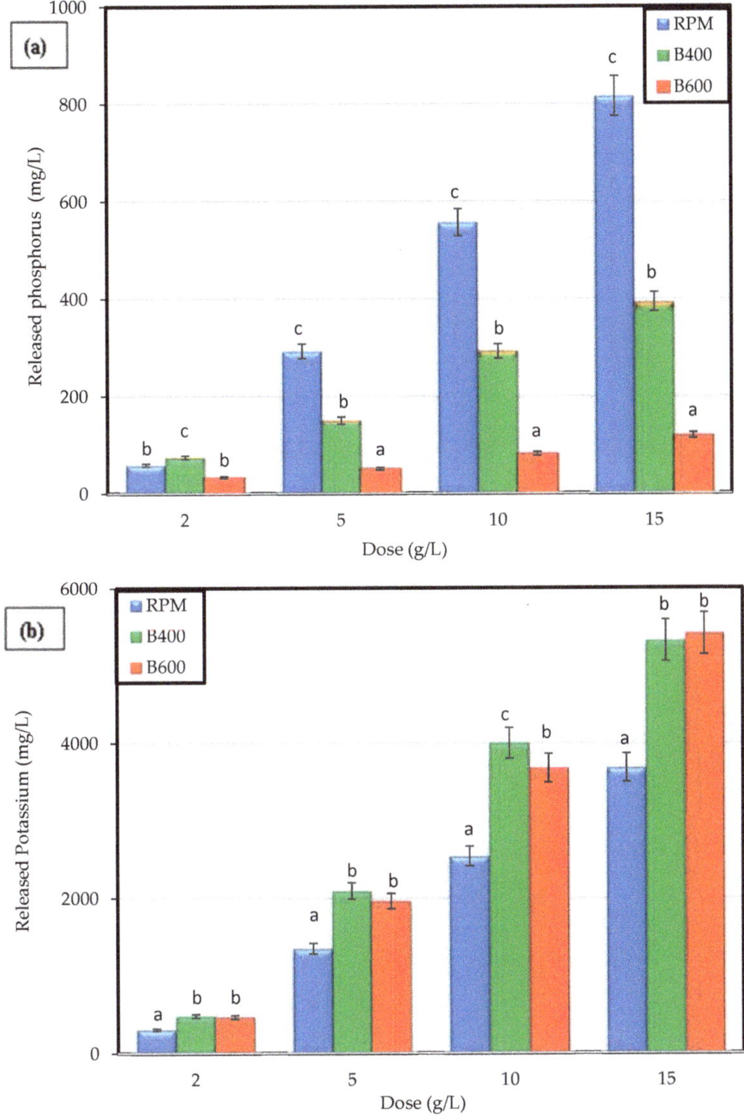

Figure 5. Effect of the used solid matrix doses on the release efficiency of phosphorus (**a**) and potassium (**b**) (pH = 6.2, T = 20 ± 2 °C). At each dose, means with the same lowercase letters are not statistically different at $p < 0.05$.

Released P or K concentrations increased significantly with the increase of the used dose. For example, P/K released concentrations increased from 58.58/304.83 to 815.47/3680.1 mg L^{-1} for RPM, from 74.45/47.88 to 393.6/5321.8 mg L^{-1} for B400, and from 33.18/464.7 to 119.13/5409.5 mg L^{-1} for B600 when the material dose increased from 2 to 15 g L^{-1}, respectively. This is mainly due to the increase of the available surface area of the solid matrices that could interact with the aqueous solutions [40]. Similar behavior was observed by Wang et al. [24] when studying the release of phosphorus from an RPM-derived biochar at 600 °C. They found that increasing the biochar concentration from 5 to 10 and then to 25 g L^{-1} resulted in an increase of the released P from 8.8 to 14.0 and 26.8 mg L^{-1}, respectively.

3.3.4. Successive Leaching Experiments—Nutrient Slow Release

The cumulative release of P and K from RPM, B400, and B600 for six successive leaching experiments (duration of 48 h for each one) is shown in Figure 6a,b, respectively.

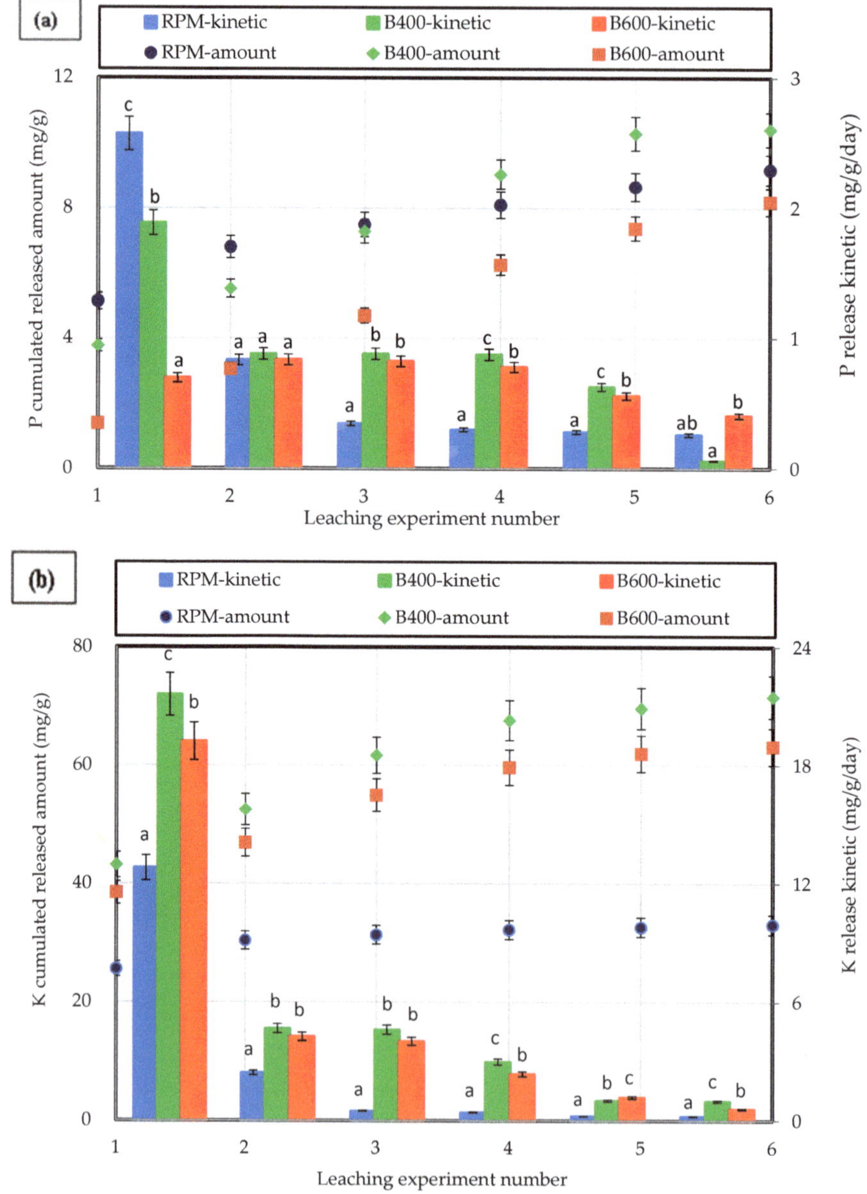

Figure 6. Effect of successive leaching experiments on the release of phosphorus (**a**) and potassium (**b**) from RPM and its two derived biochars: B400 and B600 (dosage = 10 g L-1; pH = 6.2; T = 20 ± 2 °C). At each leaching experiment number, means with the same lowercase letters are not statistically different at $p < 0.05$.

It can be clearly seen that for all the studied solid matrices, P release significantly increased with the operated number of leaching experiments. At the sixth assay, the maximal released amounts attained were approximately 9.2, 10.4, and 8.2 mg g^{-1} for RPM, B400, and B600, respectively (Figure 6a), which are 78.7%, 176.3%, and 489.5% higher than those observed in the first leaching experiment. These released amounts correspond to 54.0%, 51.2%, and 19.0% of the initial contained P in RPM, B400, and B600, respectively (Table 1). These percentages are very important compared to the ones given in the scientific literature from continuous (not successive) leaching experiments. In fact, they were assessed as only 5.5% [24] and 6.2% [22] for RPM-derived biochars collected from broiler farms in Seaford (USA) at temperatures of 400 and 600 °C and contact times of 12 and 3 days, respectively. Besides, Liang et al. [25] estimated the P release percentage to be about 10% for a biochar derived from RPM collected from a Florida dairy farm at a temperature of 450 °C and a continuous contact time of 10 days.

Compared to the first leaching experiment, cumulated P amounts considerably increased at the sixth successive leaching assay by 78.7%, 176.3%, and 489.5% for RPM, B400, and B600, respectively (Figure 6a). Concerning the kinetic release, in contrast to RPM and the fourth leaching experiment, B400 and B600 continued to release P with relatively high average release kinetic rates of 0.878 and 0.812 mg g^{-1} day^{-1}. Even after the six successive leaching experiment (12 days), B400 and B600 were still releasing P att rates of 0.062 and 0.407 mg g^{-1} day^{-1}, respectively (Figure 6a).

Similar release trends were observed for potassium. For both B400 and B600 materials, released K amounts significantly increased with the progress of the leaching experiment, attaining values of 71.6 and 63.2 mg g^{-1}, respectively, at the sixth leaching assay, corresponding to overall increase percentages of 65.7% and 64.3%, respectively, compared to the first leaching experiment (Figure 6b). These amounts represent 99.5% and 95.4% of the contained K in B400 and B600, respectively, confirming its transformation to a more leachable form and its slow release. These percentages were important when compared to continuous leaching assays where they were evaluated as only 45% to 49% for RPM-derived biochars collected from broiler farms in Seaford (USA) at temperatures varying between 400 and 600 °C, and a continuous contact time of 3 days [22].

Moreover, compared to RPM, until four successive leaching experiments, relatively high kinetic release rates of 4.071 and 3.531 mg g^{-1} day^{-1} were recorded for B400 and B600, respectively (Figure 6b). Even in the last leaching experiment, B400 and B600 were still releasing K at 0.968 and 0.587 mg g^{-1} day^{-1}, respectively (Figure 6b).

These outcomes indicate that the two RPM-derived biochars could be considered as slow-release fertilizers, where P and K could be available for several seasons in real conditions [50]. Unlike commercial fertilizers, where nutrient dissolution is very fast—even instantaneous for some products [51]—the use of RPM biochars as amendments for agricultural soils constitutes an important asset for both optimal plant growth and groundwater pollution reduction [52]. As crops also need nitrogen and other micronutrients, combining biochar application with chemical fertilizers or fresh composts has the benefits of simultaneously improving soil physical properties, nutrient retention, and plant yield [53].

4. Conclusions

This research work demonstrated that raw poultry manure pyrolysis significantly affects its physico-chemical characteristics, especially the chemical composition, pH of zero-point charge, and surface properties, including the specific surface area and functional groups. Altogether, these characteristics play an important role on the nutrient release efficiency. Nevertheless, the chemical state of these nutrients in these solid matrices, which is very dependent on the used pyrolysis temperature, might be the most important factor influencing this process. The pyrolysis process favors P conversion into a more stable phase through its probable complexation with magnesium and/or calcium and therefore reduces its release rate in the aqueous phase. In contrast, RPM pyrolysis converts potassium into a more leachable form. On the other hand, compared to RPM, RPM-derived biochars could be considered as an attractive soil conditioner with very slow nutrient release capacities. This promising property will result in a more efficient nutrient uptake by crops and limited leaching to both groundwater and surface runoff. Further investigation is currently being carried out to monitor the impact of biochar doses in agricultural soils under dynamic conditions on nutrient leaching ability and assimilation by selected plants.

Supplementary Materials: The following are available online at http://www.mdpi.com/2073-4441/11/11/2271/s1, Figure S1: SEM images of surface surfaces morphologies of RPM (a), B400 (b) and B600 (c), Figure S2: XRD images of crystalline phases of RPM, B400 and B600 (OM: organic matter; Q: quartz; C: calcite; S: sylvite and W: whitlockite), Figure S3: FTIR Spectra of RPM and its two derived biochars at temperatures of 400 and 600 °C.

Author Contributions: Conceptualization, S.H., S.J., W.K, and M.K.; methodology, S.H., S.J., J.J.L. and W.K.; software, H.H.; validation, S.J., W.K. and M.J.; formal analysis, S.H., M.J.; investigation, S.H., S.J. and W.K.; resources, M.K., J.J.L. and H.H.; data curation, S.H., S.J., M.J., H.H. and W.K.; writing—original draft preparation, S.H.; writing—review and editing, S.J., M.J., H.H. and W.K.; visualization, S.H., S.J.; supervision, S.J. and W.K.; project administration, S.J. and W.K.; funding acquisition, S.J. and W.K.

Funding: This research was funded by the Tunisian Ministry of Higher Education and Scientific Research (MHESR). The scholarship covering the mobility fees of Ms S. Hadroug from CERTE, Tunisia to UL, Ireland was taken in charge by ERASMUS+ International Credit Mobility Programme.

Acknowledgments: This research work was carried out in close collaboration between CERTE (Tunisia) and the University of Limerick (Ireland) within the framework of the ERASMUS+ International Credit Mobility Programme. The authors would like to gratefully thank the financial contribution of this Programme and the Tunisian Ministry of Higher Education and Scientific Research.

Conflicts of Interest: The authors declare no conflict of interest.

References

1. Mante, O.D.; Agblevor, F.A. Influence of pine wood shavings on the pyrolysis of poultry litter. *Waste Manag.* **2010**, *30*, 2537–2547. [CrossRef] [PubMed]

2. Akdeniz, N. A systematic review of biochar use in animal waste composting. *Waste Manag.* **2019**, *88*, 291–300. [CrossRef] [PubMed]

3. Sellami, F.; Jarboui, R.; Hachicha, S.; Medhioub, K.; Ammar, E. Co-composting of oil exhausted olive-cake, poultry manure and industrial residues of agro-food activity for soil amendment. *Bioresour. Technol.* **2008**, *99*, 1177–1188. [CrossRef] [PubMed]

4. Turan, N.G. The effects of natural zeolite on salinity level of poultry litter compost. *Bioresour. Technol.* **2008**, *99*, 2097–2101. [CrossRef] [PubMed]

5. Rodic, V.; Peric, L.; Djukic-Stojcic, M.; Vukelic, N. The environmental impact of poultry production. *Biotechnol. Anim. Husb.* **2011**, *27*, 1673–1679. [CrossRef]

6. Lu, L.; Liao, X.D.; Luo, X.G. Nutritional strategies for reducing nitrogen, phosphorus and trace mineral excretions of livestock and poultry. *J. Integr. Agric.* **2017**, *16*, 2815–2833. [CrossRef]

7. Hu, Y.; Cheng, H.; Tao, S. Environmental and human health challenges of industrial livestock and poultry farming in China and their mitigation. *Environ. Int.* **2017**, *107*, 111–130. [CrossRef]

8. Cely, P.; Gascó, G.; Paz-Ferreiro, J.; Méndez, A. Agronomic properties of biochars from different manure wastes. *J. Anal. Appl. Pyrolysis* **2015**, *111*, 173–182. [CrossRef]

9. Lynch, D.; Henihan, A.M.; Kwapinski, W.; Zhang, L.; Leahy, J.J. Ash agglomeration and deposition during combustion of poultry litter in a bubbling fluidized-bed combustor. *Energy Fuels* **2013**, *27*, 4684–4694. [CrossRef]

10. Taupe, N.C.; Lynch, D.; Wnetrzak, R.; Kwapinska, M.; Kwapinski, W.; Leahy, J.J. Updraft gasification of poultry litter at farm-scale—A case study. *Waste Manag.* **2016**, *50*, 324–333. [CrossRef]

11. Kelleher, B.P.; O'Callaghan, M.N.; Leahy, M.J.; O'Dwyer, T.F.; Leahy, J.J. The use of fly ash from the combustion of poultry litter for the adsorption of chromium (III) from aqueous solution. *J. Chem. Technol. Biotechnol.* **2002**, *77*, 1212–1218. [CrossRef]

12. Agyarko-Mintah, E.; Cowie, A.; Singh, B.P.; Joseph, S.; Van Zwieten, L.; Cowie, A.; Harden, S.; Smillie, R. Biochar increases nitrogen retention and lowers greenhouse gas emissions when added to composting poultry litter. *Waste Manag.* **2017**, *61*, 138–149. [CrossRef] [PubMed]

13. Wongrod, S.; Simon, S.; Guibaud, G.; Lens, P.N.L.; Pechaud, Y.; Huguenot, D.; van Hullebusch, E.D. Lead sorption by biochar produced from digestates: Consequences of chemical modification and washing. *J. Environ. Manag.* **2018**, *219*, 277–284. [CrossRef]

14. Kwapinski, W.; Byrne, C.M.P.; Kryachko, E.; Wolfram, P.; Adley, C.; Leahy, J.J.; Novotny, E.H.; Hayes, M.H.B. Biochar from biomass and waste. *Waste Biomass Valoriz.* **2010**, *1*, 177–189. [CrossRef]

15. Mau, V.; Gross, A. Energy conversion and gas emissions from production and combustion of poultry-litter-derived hydrochar and biochar. *Appl. Energy* **2018**, *213*, 510–519. [CrossRef]

16. Hu, X.; Gholizadeh, M. Biomass Pyrolysis: A Review of the Process Development and Challenges from Initial Researches up to the Commercialisation Stage. *J. Energy Chem.* **2019**, 109–143. [CrossRef]

17. Ding, Y.; Liu, Y.; Liu, S.; Huang, X.; Li, Z.; Tan, X.; Zeng, G.; Zhou, L. Potential Benefits of Biochar in Agricultural Soils: A Review. *Pedosphere* **2017**, *27*, 645–661. [CrossRef]

18. Shashvatt, U.; Benoit, J.; Aris, H.; Blaney, L. CO_2-assisted phosphorus extraction from poultry litter and selective recovery of struvite and potassium struvite. *Water Res.* **2018**, *143*, 19–27. [CrossRef]

19. Ghanim, B.M.; Pandey, D.S.; Kwapinski, W.; Leahy, J.J. Hydrothermal carbonisation of poultry litter: Effects of treatment temperature and residence time on yields and chemical properties of hydrochars. *Bioresour. Technol.* **2016**, *216*, 373–380. [CrossRef]

20. Azzaz, A.A.; Jellali, S.; Akrout, H.; Assadi, A.A.; Bousselmi, L. Optimization of a cationic dye removal by a chemically modified agriculture by-product using response surface methodology: Biomasses characterization and adsorption properties. *Environ. Sci. Pollut. Res.* **2016**, *24*, 9831–9846. [CrossRef]

21. Abdallah, M.M.; Ahmad, M.N.; Walker, G.; Leahy, J.J.; Kwapinski, W. Batch and Continuous Systems for Zn, Cu, and Pb Metal Ions Adsorption on Spent Mushroom Compost Biochar. *Ind. Eng. Chem. Res.* **2019**, *58*, 7296–7307. [CrossRef]

22. Song, W.; Guo, M. Quality variations of poultry litter biochar generated at different pyrolysis temperatures. *J. Anal. Appl. Pyrolysis* **2012**, *94*, 138–145. [CrossRef]

23. Trazzi, P.A.; Leahy, J.J.; Hayes, M.H.B.; Kwapinski, W. Adsorption and desorption of phosphate on biochars. *J. Environ. Chem. Eng.* **2016**, *4*, 37–46. [CrossRef]

24. Wang, Y.; Lin, Y.; Chiu, P.C.; Imhoff, P.T.; Guo, M. Phosphorus release behaviors of poultry litter biochar as a soil amendment. *Sci. Total Environ.* **2015**, *512–513*, 454–463. [CrossRef] [PubMed]

25. Liang, Y.; Cao, X.; Zhao, L.; Xu, X.; Harris, W. Phosphorus Release from Dairy Manure, the Manure-Derived Biochar, and Their Amended Soil: Effects of Phosphorus Nature and Soil Property. *J. Environ. Qual.* **2014**, *43*, 1504–1509. [CrossRef]

26. Azzaz, A.A.; Jellali, S.; Assadi, A.A.; Bousselmi, L. Chemical treatment of orange tree sawdust for a cationic dye enhancement removal from aqueous solutions: Kinetic, equilibrium and thermodynamic studies. *Desalin. Water Treat.* **2016**, *57*, 22107–22119. [CrossRef]

27. Haddad, K.; Jellali, S.; Jeguirim, M.; Ben Hassen Trabelsi, A.; Limousy, L. Investigations on phosphorus recovery from aqueous solutions by biochars derived from magnesium-pretreated cypress sawdust. *J. Environ. Manag.* **2018**, *216*, 305–314. [CrossRef]

28. Xu, Y.; Chen, B. Investigation of thermodynamic parameters in the pyrolysis conversion of biomass and manure to biochars using thermogravimetric analysis. *Bioresour. Technol.* **2013**, *146*, 485–493. [CrossRef]

29. Zhao, L.; Cao, X.; Mašek, O.; Zimmerman, A. Heterogeneity of biochar properties as a function of feedstock sources and production temperatures. *J. Hazard. Mater.* **2013**, *256–257*, 1–9. [CrossRef]

30. Cimo, G.; Kucerik, J.; Berns, A.E.; Schaumann, G.E.; Alonzo, G.; Conte, P. Effect of Heating Time and Temperature on the Chemical Characteristics of Biochar from Poultry Manure. *J. Agric. Food Chem.* **2014**, *62*, 1912–1918. [CrossRef]

31. Bekiaris, G.; Peltre, C.; Jensen, L.S.; Bruun, S. Using FTIR-photoacoustic spectroscopy for phosphorus speciation analysis of biochars. *Spectrochim. Acta Part A Mol. Biomol. Spectrosc.* **2016**, *168*, 29–36. [CrossRef] [PubMed]

32. Kalderis, D.; Hawthorne, S.B.; Clifford, A.A.; Gidarakos, E. Interaction of soil, water and TNT during degradation of TNT on contaminated soil using subcritical water. *J. Hazard. Mater.* **2008**, *159*, 329–334. [CrossRef] [PubMed]

33. Jindo, K.; Mizumoto, H.; Sawada, Y.; Sanchez-Monedero, M.A.; Sonoki, T. Physical and chemical characterization of biochars derived from different agricultural residues. *Biogeosciences* **2014**, *11*, 6613–6621. [CrossRef]

34. Haddad, K.; Jellali, S.; Jaouadi, S.; Benltifa, M.; Mlayah, A.; Hamzaoui, A.H. Raw and treated marble wastes reuse as low cost materials for phosphorus removal from aqueous solutions: Efficiencies and mechanisms. *C. R. Chim.* **2015**, *18*, 75–87. [CrossRef]

35. Cao, X.; Harris, W. Properties of dairy-manure-derived biochar pertinent to its potential use in remediation. *Bioresour. Technol.* **2010**, *101*, 5222–5228. [CrossRef]

36. Wang, M.; Zhu, Y.; Cheng, L.; Andserson, B.; Zhao, X.; Wang, D.; Ding, A. ScienceDirect Review on utilization of biochar for metal-contaminated soil and sediment remediation. *J. Environ. Sci.* **2017**, *63*, 156–173. [CrossRef]

37. Novais, S.V.; Zenero, M.D.O.; Barreto, M.S.C.; Montes, C.R.; Cerri, C.E.P. Phosphorus removal from eutrophic water using modified biochar. *Sci. Total Environ.* **2018**, *633*, 825–835. [CrossRef]

38. Novak, J.M.; Busscher, W.J.; Laird, D.L.; Ahmedna, M.; Watts, D.W.; Niandou, M.A.S. Impact of Biochar Amendment on Fertility of a Southeastern Coastal Plain Soil. *Soil Sci.* **2009**, *174*, 105–112. [CrossRef]

39. Cantrell, K.B.; Hunt, P.G.; Uchimiya, M.; Novak, J.M.; Ro, K.S. Impact of pyrolysis temperature and manure source on physicochemical characteristics of biochar. *Bioresour. Technol.* **2012**, *107*, 419–428. [CrossRef]

40. Chen, X.; Lin, Q.; He, R.; Zhao, X.; Li, G. Hydrochar production from watermelon peel by hydrothermal carbonization. *Bioresour. Technol.* **2017**, *241*, 236–243. [CrossRef]

41. Domingues, R.R.; Trugilho, P.F.; Silva, C.A.; De Melo, I.C.N.A.; Melo, L.C.A.; Magriotis, Z.M.; Sánchez-Monedero, M.A. Properties of biochar derived from wood and high-nutrient biomasses with the aim of agronomic and environmental benefits. *PLoS ONE* **2017**, *12*, e0176884. [CrossRef] [PubMed]

42. Enders, A.; Hanley, K.; Whitman, T.; Joseph, S.; Lehmann, J. Characterization of biochars to evaluate recalcitrance and agronomic performance. *Bioresour. Technol.* **2012**, *114*, 644–653. [CrossRef] [PubMed]

43. Cao, W.; Cao, C.; Guo, L.; Jin, H.; Dargusch, M.; Bernhardt, D.; Yao, X. Hydrogen production from supercritical water gasification of chicken manure. *Int. J. Hydrogen Energy* **2016**, *41*, 22722–22731. [CrossRef]

44. Mau, V.; Quance, J.; Posmanik, R.; Gross, A. Phases' characteristics of poultry litter hydrothermal carbonization under a range of process parameters. *Bioresour. Technol.* **2016**, *219*, 632–642. [CrossRef] [PubMed]

45. Sun, J.; Hoon, S.; Jung, S.; Ryu, C.; Jeon, J.; Shin, M.; Park, Y. Production and utilization of biochar: A review. *J. Ind. Eng. Chem.* **2016**, *40*, 1–15.

46. Xiao, R.; Wang, J.J.; Gaston, L.A.; Zhou, B.; Park, J.H.; Li, R.; Dodla, S.K.; Zhang, Z. Biochar produced from mineral salt-impregnated chicken manure: Fertility properties and potential for carbon sequestration. *Waste Manag.* **2018**, *78*, 802–810. [CrossRef]

47. Halajnia, A.; Oustan, S.; Najafi, N.; Khataee, A.R.; Lakzian, A. Adsorption-desorption characteristics of nitrate, phosphate and sulfate on Mg-Al layered double hydroxide. *Appl. Clay Sci.* **2013**, *80–81*, 305–312. [CrossRef]

48. Peak, D.; Sims, J.T.; Sparks, D.L. Solid-state speciation of natural and alum-amended poultry litter using XANES spectroscopy. *Environ. Sci. Technol.* **2002**, *36*, 4253–4261. [CrossRef]

49. Barca, C.; Gérente, C.; Meyer, D.; Chazarenc, F.; Andrès, Y. Phosphate removal from synthetic and real wastewater using steel slags produced in Europe. *Water Res.* **2012**, *46*, 2376–2384. [CrossRef]

50. Mukherjee, A.; Zimmerman, A.R. Organic carbon and nutrient release from a range of laboratory-produced biochars and biochar-soil mixtures. *Geoderma* **2013**, *193*, 122–130. [CrossRef]

51. Liu, G.; Zotarelli, L.; Li, Y.; Dinkins, D.; Wang, Q. *Controlled-Release and Slow-Release Fertilizers as Nutrients Management Tools*; UF/IFAS Extension: Gainesville, FL, USA, 2017.

52. Sun, K.; Qiu, M.; Han, L.; Jin, J.; Wang, Z.; Pan, Z.; Xing, B. Speciation of phosphorus in plant- and manure-derived biochars and its dissolution under various aqueous conditions. *Sci. Total Environ.* **2018**, *634*, 1300–1307. [CrossRef] [PubMed]

53. Schulz, H.; Glaser, B. Effects of biochar compared to organic and inorganic fertilizers on soil quality and plant growth in a greenhouse experiment. *J. Plant Nutr. Soil Sci.* **2012**, *175*, 410–422. [CrossRef]

Article

Optimization of Green Extraction and Purification of PHA Produced by Mixed Microbial Cultures from Sludge

Guilherme A. de Souza Reis [1,*, Michiel H. A. Michels [1], Gabriela L. Fajardo [1,2], Ischa Lamot [1] and Jappe H. de Best [1]**

[1]   Centre of Expertise Biobased Economy, Avans University of Applied Sciences, Lovensdijkstraat 63, 4818 AJ Breda, The Netherlands; mha.michels@avans.nl (M.H.A.M.); gabrielalfajardo@gmail.com (G.L.F.); i.lamot@avans.nl (I.L.); jh.debest@avans.nl (J.H.d.B.)
[2]   Department of Chemical Engineering, Universidade Federal de Minas Gerais, Belo Horizonte 31270-901, Brazil
*   Correspondence: guilhermeasreis@gmail.com; Tel.: +31-(0)651924343

Received: 31 March 2020; Accepted: 17 April 2020; Published: 21 April 2020

Abstract: Sludge from municipal wastewater treatment systems can be used as a source of mixed microbial cultures for the production of polyhydroxyalkanoates (PHA). Stored intracellularly, the PHA is accumulated by some species of bacteria as energy stockpile and can be extracted from the cells by reflux extraction. Dimethyl carbonate was tested as a solvent for the PHA extraction at different extraction times and biomass to solvent ratios, and 1-butanol was tested for purifying the obtained PHA at different purification times and PHA to solvent ratios. Overall, only a very small difference was observed in the different extraction scenarios. An average extraction amount of 30.7 ± 1.6 g of PHA per 100 g of biomass was achieved. After purification with 1-butanol, a visual difference was observed in the PHA between the tested scenarios, although the actual purity of the resulting samples did not present a significant difference. The overall purity increased from 91.2 ± 0.1% to 98.0 ± 0.1%.

Keywords: polyhydroxyalkanoates; PHA; PHBV; mixed microbial culture; green extraction; dimethyl carbonate; purification; 1-butanol; wastewater valorization

---

1. Introduction

The constant search for environmentally-friendly alternatives for fossil-based materials has been backed up lately by an increase of research in the field of bioplastics such as polyhydroxyalkanoates (PHAs). PHAs are biodegradable polymers synthesized by a variety of bacteria in intracellular granules that serve as energy storage [1,2]. Industrial PHA production has been made feasible by using selected strains of pure microbial cultures to ferment refined substrates [3]. However, a much more sustainable, and perhaps cheaper, option can be found in the use of industrial residue streams as a source of bacterial feed [4–6]. VFAs are organic acids with an aliphatic chain of less than five carbons which can be present in or derived from a large variety of residue streams. VFAs have been shown to be an interesting and very feasible feedstock in the PHA production process by both pure and mixed microbial cultures [7].

Applying a mixed microbial culture (MMC) in the process could furthermore reduce the costs of PHA production, because sterilization of the substrate and reactors is not needed. It has been observed that activated sludge of municipal wastewater treatment plants can be used as a source of MMC with a good PHA-accumulating potential [8,9]. Certain fermentation strategies can be used to explore this accumulating potential and generate a PHA-rich biomass [10–12]. A very useful method is a dynamic fed-batch fermentation with alternating repeated periods of feast and famine, which

---

can also be combined with an aerobic or anaerobic environment [13–15]. Through application of a pulsed VFA-feeding regime, it is possible to reduce the effects of too extreme pH variations caused by the addition of VFAs to the medium, as well as to favor the maintenance of the PHA-accumulating bacterial population over other non-accumulating species during the famine periods. This method also stimulates the PHA-accumulating bacteria to stockpile the biopolymer intracellularly throughout multiple feast and famine cycles, which highlights this feed-on-demand process amongst other approaches even on an industrial scale [8].

When studying the use of different feedstocks for PHA production, it is also important to understand the relationship between the feed composition and the monomeric proportions of the resulting polymer. When VFAs are the bacterial feed, acetic and propionic acid are the main precursors in a mechanism for the production of the monomers 3-hydroxybutyrate (3HB) and 3-hydroxyvalerate (3HV) in PHA [16–19]. Acetic acid can be converted into 3HB via acetyl-CoA, while acetic acid together with the odd numbered propionic acid are used to form 3HV via the conversion to acetyl-CoA and propionyl-CoA [17,20]. The combined production of 3HB and 3HV as monomers leads to the synthesis of copolymer poly(3-hydroxybutyrate-co-3-hydroxyvalerate) (PHBV) (Figure 1).

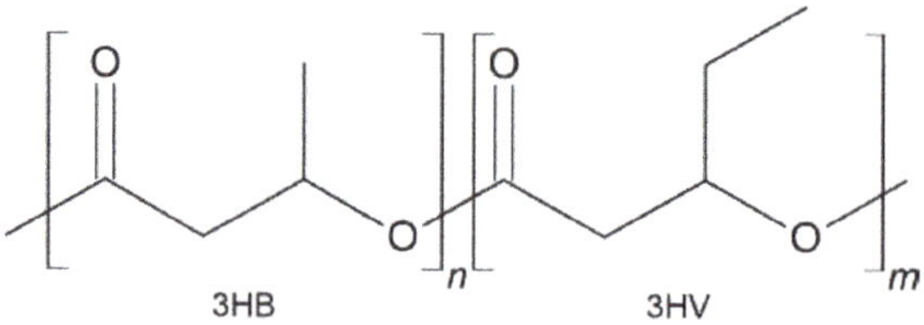

Figure 1. Representation of molecular structure of poly (3-hydroxybutyrate-co-3-hydroxyvalerate) (PHBV) (Drawn on ACD/ChemSketch Freeware, version 2019.2.1, Advanced Chemistry Development, Inc., Toronto, ON, Canada, www.acdlabs.com, 2020).

Extraction is the next step in the production process of PHA, which is done by solubilizing the intracellular PHA followed by separation of the extracted residual biomass and isolation of PHA from the solvent. Reflux and Soxhlet extractions, with and without biomass pretreatments, have been described in literature for many different solvents [21–24]. Non-halogenated solvents have been the focus of many researches for their reduced toxicity, although the chlorinated ones, such as chloroform and dichloromethane, are still considered reference solvents because of their high efficiency [22,25,26]. The non-halogenated solvent dimethyl carbonate performs much better than a range of solvents such as diethyl carbonate, propylene carbonate and ethyl acetate and it achieved satisfactory yields of PHA recovery when compared to dichloromethane [27,28]. Furthermore, dimethyl carbonate (DMC) is considered to be a green solvent for its low toxicity [29] when compared to chloroform and dichloromethane [30]. Therefore, DMC was chosen as the solvent for PHA extraction in the present work. 1-Butanol has also been shown to be efficient as a solvent for PHA extraction [9], with the advantage of leading to a simple separation process through gelation of the polymer when cooling down the mixture [31]. Due to this easiness in separating the solvent from the solid PHA, 1-butanol was chosen as a purification agent in this study.

Total PHA content in biomass and purity evaluation of the obtained polymer can both be done with various simple analytical techniques. Thermal gravimetric analysis (TGA) [32] can be applied for a quick overview of these parameters when focusing on the degradation temperature of the produced PHA. Gas chromatography combined with mass spectrometry (GC-MS) is a more accurate technique that allows the investigation of the monomer concentration, composition and purity of the product [33]. Some studies made a parallel in between these techniques, showing that, even though different PHA content values were obtained by each method, there is a direct correlation between them [34]. For the GC-MS analysis, a pre-treatment step has to be added for the PHA to be able to be analyzed.

In the present work, the aim was to optimize a green extraction and purification of PHA from a mixed microbial culture, using dimethyl carbonate and 1-butanol, respectively. The biomass to solvent ratio or PHA to solvent ratio and the extraction or purification time were the parameters to be optimized.

2. Material and Methods

2.1. Fermentation Process for PHA Production

The fermentation process was adapted from the work of Valentino (2015) and the patent of Werker (2013) [8,12]. The accumulation of PHA was done in two identical 2.5 L bioreactors (Infors™ MINI-2.5-BACT, Bottmingen, Switzerland) set at 25 °C and 200 rpm and aerated with 1.5 L min^{-1} of air. Thickened secondary sludge from the wastewater treatment plant of water board Brabantse Delta in Bath, the Netherlands, was used as a source of PHA-accumulating bacteria. A synthetic feed was made with a solution of acetic and propionic acid in a molar proportion of 3:1, respectively, and a total chemical oxygen demand (COD) of 20 g L^{-1}. This feed composition guarantees that the PHA produced is the co-polymer PHBV in an expected monomer distribution of 50% of each [17]. To start the process, 500 mL of sludge with volatile suspended solids (VSS) concentration of about 8 g L^{-1} was mixed with 500 mL of tap water in the reactor. The mixture was then left for 2 h without any feed under the fermentation conditions in order to stabilize the biomass. A single feed pulse of 10 mL was then given, followed by the first starvation period of 1 hour. By the end of this period, the dissolved oxygen (DO) level was then considered to be the maximum achieved. The feed-on-demand was then automatically controlled by a DO value corresponding to 80% of the maximum as the set point as the condition for the next pulse feed. The feed pulses corresponded to a volume of 10 mL given over 1 min every time the mentioned condition was met. The whole process was set to a total of 22 h. The process was automatically stopped by dosing hydrochloric acid until a pH of at least 2 was reached in order to cease the bacterial metabolism. At that point, stirring and air inlet were shut down.

2.2. Homogenization of Biomass

The mixture inside the bioreactors was left for a few minutes to settle and the PHA-rich biomass was centrifuged (5810 series, Eppendorf™, Nijmegen, The Netherlands) and washed twice with tap water. The biomass was then freeze-dried (CHRIST Alpha 1-4 LD Plus) overnight. The dried biomass resulted from 20 runs (10 in each bioreactor) was mixed and made homogeneous using a mortar and pestle.

2.3. Extraction Process

The PHA was extracted from the biomass via reflux by adding 25 mL of dimethyl carbonate (DMC) as a solvent to different amounts of biomass in a round-bottom flask connected to a cooling column. Different biomass to solvent ratios, 0.01 g mL^{-1}, 0.025 g mL^{-1}, 0.05 g mL^{-1}, and 0.1 g mL^{-1}, were tested in duplicate by adjusting the amount of biomass. These biomass to solvent ratios will be further referred as 1%, 2.5%, 5%, and 10%, respectively. The round-bottom flask was immersed in a pan filled with glycerin, previously heated up to the boiling point of the solvent (90 °C for the DMC). Different times of extraction were tested (0.5 h, 1 h, 1.5 h, and 2 h). After the extraction, the pan was removed to allow the flask to naturally cool down to room temperature. A vacuum filtration with Whatman™ paper filters was then used to separate the biomass from the solution. A rotary evaporator (Hei-VAP Value, Heidolph™, Schwabach, Germany) was used to recover the solvent and to separate the PHA in the form of a film attached to the wall of the flask. The obtained PHA was left to dry overnight and then weighed. The experimental data obtained under the different extraction conditions was compared using first an F-test for variances and then the adequate t-test for the p values. Chloroform and dichloromethane were used as reference extraction solvents [26–28], for which a biomass to solvent ratio of 1% and an extraction time of 1 h were used.

2.4. Purification Process

The extracted PHA was purified with 1-butanol (≥99%, Sigma-Aldrich™, Zwijndrecht, The Netherlands) via reflux. 20 mL of the solvent was added to different amounts of the polymer in a round-bottom flask which was connected to a cooling column. The different PHA to solvent ratios tested in duplicate were 0.01 g mL^{-1}, 0.02 g mL^{-1}, and 0.04 g mL^{-1} (1, 2 and 4%). The flask was immersed in a pan with glycerin and heated up to the boiling point of 1-butanol (117.7 °C). The purification times tested were 0.5 h, 1 h, and 2 h. The round-bottom flask was then taken from the pan, closed with a cap and allowed to cool down at room temperature overnight. The solvent was then separated from the jelly-like polymer by physically pressing it out [31] with a piece of cloth. The PHA was allowed to dry at room temperature overnight. The 1-butanol was separated from the solubilized contaminants and recycled with a rotary evaporator (Hei-VAP Value, Heidolph™, Schwabach, Germany). The experimental data obtained under the different purification conditions was compared using first an F-test for variances and then the adequate t-test for the p values.

2.5. Analytical Methods

2.5.1. Thermal Gravimetric Analysis (TGA)

Samples of produced biomass were analyzed in duplicate for PHA content and every sample of extracted and purified PHA was analyzed for purity with TGA (TGA 500Q, TA instruments, Etten-Leur, The Netherlands). Because of the feed composition, all PHA produced was considered to be the co-polymer PHBV. A mass of around 10 mg was placed into platinum pans and a ramp mode of 5 °C per minute until 600 °C was set under nitrogen atmosphere, with flow rates of 40 mL min^{-1} on the balance, 60 mL min^{-1} on the sample (adapted from Hahn and Chang, 1995) [34]. The mass of PHA in each sample was determined as the mass loss in the temperature range between 250 °C and 270 °C. A commercial sample of PHBV was used to establish this range.

2.5.2. Gas Chromatography/Mass Spectrometry (GC-MS)

Sample Preparation

Following the procedure in Lo et al. (2009) [33], an acidic methanolysis reaction was used to hydrolyze the polymeric chain and to convert the monomers into their methylated ester form. A PHA sample of each experiment was weighed for a mass between 2 and 5 mg and brought into a 5 mL reaction vial. 1 mL of chloroform (99%, Sigma-Aldrich™, Zwijndrecht, The Netherlands), 0.95 mL of methanol (Technical grade, BOOM™, Meppel, The Netherlands) and 0.05 mL of sulfuric acid (95%–98%, Sigma-Aldrich™, Zwijndrecht, The Netherlands) were added into the reaction vial which was closed tightly and shaken. The reaction vial was placed into a heating block (Techne Dri-block™, Staffordshire, England, UK) at 100 °C for 6 h and shaken every 1 h. The vial was then left to cool down to room temperature. The solution was transferred from the vial to a centrifuge tube and 1 mL of 1 M NaCl (Extra pure, BOOM™, Meppel, The Netherlands) solution was added. The tube was shaken and centrifuged (5810 series, Eppendorf™, Nijmegen, The Netherlands) for 3 minutes at 4000 rpm. The aqueous layer was discarded and another 1 mL of the NaCl solution was added. The tube was shaken and centrifuged again. The organic phase was then taken from the tube with a needle attached to a 1 mL syringe. The solution volume was measured with the syringe graduation, filtered (0.45 μm × 13 mm PTFE filter) and transferred to a vial.

GC-MS Settings

A calibration curve was made using standards of methyl-(R)-3-hydroxybutyrate and methyl-(R)-3-hydroxyvalerate (>98%, Sigma-Aldrich™, Zwijndrecht, The Netherlands) using chloroform as a solvent to achieve different dilutions. The GC-MS used was a 7820A GC System/5977E MSD Agilent Technologies™ with a HP-5MS capillary GC column (30 m × 0.25 mm, 0.25 μm), flow of

2 mL min^{-1} helium, sample injection of 1 μL, temperature of 250 °C at the injector and detector with a heating rate of 10 °C min^{-1}. The software NIST MS Search Program was used to identify the samples components through their mass spectrum.

3. Results and Discussion

3.1. PHA Accumulation

Although there were slight variations throughout the runs, the fermentation process was consistent, with an initial volatile suspended solids concentration of about 4 g L^{-1}. A range of 23 to 25 pulse feeds were given during each of the 20 runs. Each fermentation run resulted in around 5.5 g of dry biomass. It was observed that the biomass easily settled to the bottom of the vessel.

The TGA analysis of the dry biomass revealed a content of around 40% of PHA in mass, represented by a degradation peak at the temperature range of 265 to 277 °C, which was proved to be the right degradation temperature by comparing the analysis with a commercial sample of PHBV. Hahn and Chang (1995) [34] discovered a correlation between the PHA content measured through TGA and the PHA content measure through GC analysis, where the result from GC analysis are considered to be more accurate. This correlation is expressed as a linear model:

$$y = 1.16x - 15.27 \tag{1}$$

where x is the PHA content by TGA and y the real content. Using this correlation, the total PHA content is around 32%. This result is slightly lower, but still close, to the ones mentioned in the PHARIO report [9], for which the same MMC source was used and the PHA accumulation resulted in values around 39 g of PHA per 100 g of VSS. A difference, however, that might explain the higher production result in that work is the nitrogen and phosphorous supplement in the feed composition and the extra feeds during acclimation process of the biomass.

3.2. Extraction

The results of PHA extraction in all studied conditions were around 31 g of PHA per 100 g of biomass, with slightly higher extraction values at lower biomass to solvent ratios and longer extraction times (Table 1). This result indicates a very high polymer recovery, which contrasts with what is discussed in Samorì (2015) [27], where only about half of all the PHA inside the MMC biomass could be extracted with dimethyl carbonate without any cell pretreatment. It is important to mention, however, that the MMC used in that work for PHA accumulation has a different source and it was submitted to an extensive process of bacterial selection over time, which might affect the general composition of the biomass and, perhaps, the efficiency of the DMC as a solvent for PHA extraction.

Table 1. Extraction results of PHA with dimethyl carbonate. Freeze dried biomass after polyhydroxyalkanoates (PHA) accumulation was extracted in 25 mL dimethyl carbonate at its boiling point at different biomass to solvent ratios for different extraction times. The biomass to solvent ratio is expressed as a percentage of grams of biomass per 100 ml of solvent. Values represent averages ± standard deviation of duplicates in g of PHA per 100 g of biomass.

Biomass to Solvent Ratio (%)	0.25 h	0.5 h	1 h	1.5 h	2 h
1%	31.4 ± 1.0[abc]	32.2 ± 0.1[b]	31.7 ± 0.2[bd]	32.9 ± 0.2[b]	
2.5%	30.6 ± 0.4[cd]	30.4 ± 0.3[cd]	31.5 ± 0.1[ad]	32.3 ± 0.4[bd]	
5%		30.5 ± 0.1[c]	30.0 ± 1.0[abc]	30.8 ± 0.7[abc]	30.4 ± 1.3[abc]
10%			29.3 ± 1.1[abc]	27.8 ± 0.4[a]	28.2 ± 0.5[ac]

[abcd] Average values not sharing a common superscript were significantly different ($p < 0.05$).

The scenarios with a 10% biomass to solvent ratio presented some practical issues because of the relatively high amount of biomass that settled in the bottom part of the extraction flask in direct contact with the heating source and with low or no contact to the solvent.

The PHA extraction process with DMC was compared with chloroform [23] and dichloromethane [28] as the reference solvents (Table 2).

Table 2. Comparative extraction results for the different solvents. The reflux extraction of PHA with dimethyl carbonate at 1% biomass to solvent ratio for 1 h was compared with the reflux extraction with chloroform and with dichloromethane at the same solvent ratio and time. Values represent averages ± standard deviation of duplicates of g of PHA per 100 g of biomass.

Solvent	g PHA/100 g Biomass
Dimethyl carbonate	31.7 ± 0.2
Chloroform	37.5 ± 0.2
Dichloromethane	39.0 ± 0.2

Although the amount of extracted PHA seems higher with chloroform or dichloromethane compared with dimethyl carbonate as a solvent, when manually stretched, the polymer films produced with chloroform and dichloromethane were both very brittle and not much elastic. They would immediately break apart when pulled. The brittleness was caused by a higher percentage of impurities in the PHA extracted with chloroform and dichloromethane, as it is further discussed in the results obtained with TGA. The PHA plastic films produced in the process with dimethyl carbonate, on the other hand, had a much higher plastic deformation capability, similar to a common strong plastic bag. However, regardless of the solvent used, the resulting solid PHA had a green/brown color after the solvent recovery in all the produced samples (Figure 2).

Figure 2. A sample of PHA film produced directly after the solvent recovery. The dark green color was common to all of the samples produced.

A TGA of PHA samples obtained directly by extraction with dimethyl carbonate reveals a purity of 91.2 ± 0.1% versus a purity of 82.5 ± 3.3% for the extraction with chloroform and of 86.4% ± 3.7% with dichloromethane. These results could explain the higher yields obtained for the extraction process with chloroform and dichloromethane meaning that these reference solvents are solubilizing not only the PHA, but also higher amounts of other compounds present in the biomass, which results in lower overall purity in these samples.

3.3. Purification

Purification of the extracted PHA with 1-butanol revealed that the whitest product was obtained with a PHA to solvent ratio of 1% and after 0.5 h of purification time (Figure 3).

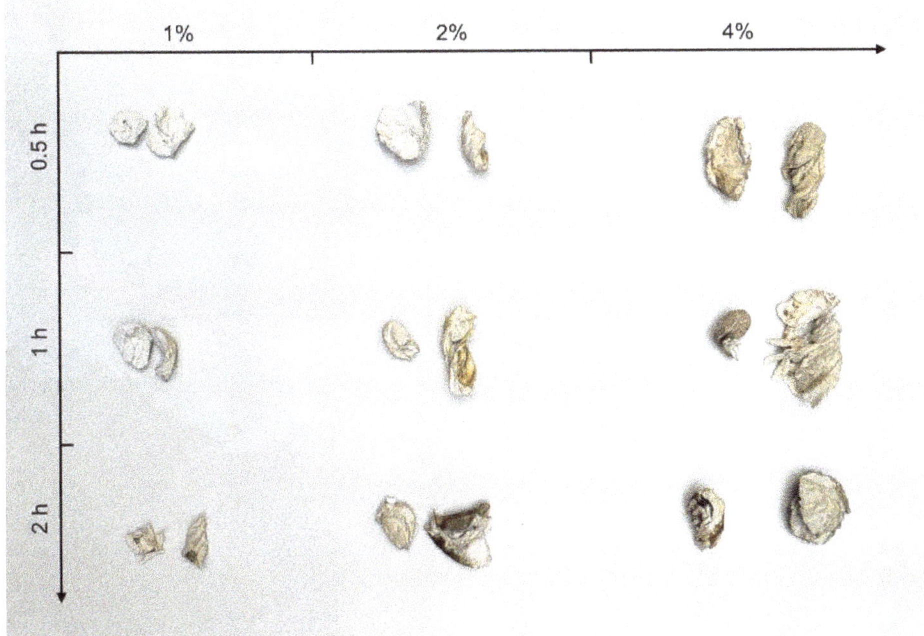

Figure 3. Comparison between the purified PHA at different experimental points.

It was expected that a lower PHA to solvent ratio led to PHA with less impurities. However, a longer purification time led to a darker-colored product, although not much difference was registered in the actual purity of the samples (Table 3).

Table 3. Purity results by TGA of the samples submitted to the purification process by 1-butanol. The PHA to solvent ratio is written as in grams of PHA per 100 ml of solvent Values represent averages ± standard deviation of triplicates of percentage of purity of purified PHA.

PHA to Solvent Ratio (%)	0.5 h	1 h	2 h
1%	98.3 ± 0.1^{acd}	98.5 ± 0.1^{a}	98.2 ± 0.2^{acd}
2%	98.4 ± 0.1^{bc}	98.9 ± 0.1^{ac}	97.9 ± 0.4^{acd}
4%	98.3 ± 0.3^{acd}	98.2 ± 0.1^{bd}	98.4 ± 0.3^{acd}

[abcd] Average values not sharing a common superscript were significantly different ($p < 0.05$).

The evaluation of PHA by TGA revealed an increase in purity from $91.2 \pm 0.1\%$ to $98.0 \pm 0.1\%$ after purification with 1-butanol (Figure 4). The peak degradation temperature of the PHA was identified to be $253.4 \pm 7.3\ °C$ which is comparable to literature about different monomer compositions of the PHBV copolymer [35–37].

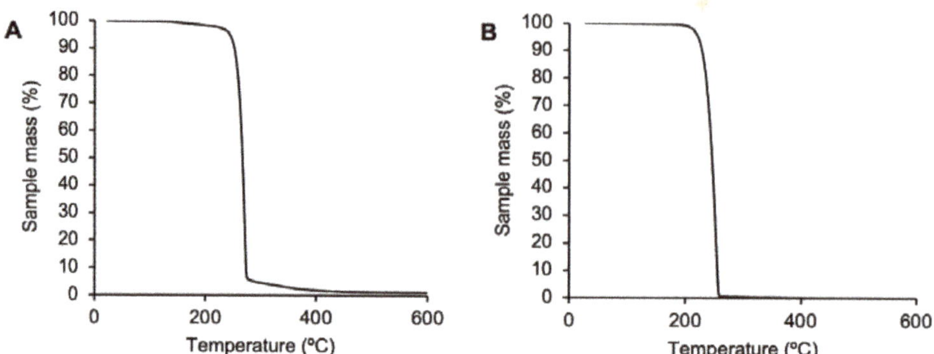

Figure 4. Thermal gravimetric analysis (TGA) of extracted PHA before (**A**) and after (**B**) purification with 1-butanol.

3.4. Analysis by Gas Chromatography-Mass Spectrometry (GC-MS)

Samples of the extracted PHA before and after purification were analyzed with GC-MS for its monomeric composition and identification of impurities (Figure 5). Besides 3-hydroxybutyrate and 3-hydroxyvalerate, the monomer 3-hydroxy-2-methylvalerate was also present in minor quantities in the samples. This monomer has been reported already as a common component of polymers synthesized by enriched cultures of glycogen-accumulating organisms (GAO) [35,38,39].

The GC-MS analysis revealed a monomer composition of 35.6 ± 2.5% 3-hydroxybutyrate and 64.4 ± 2.5% 3-hydroxyvalerate. Given the feed composition, a monomer distribution of the produced PHA of 50% 3-hydroxybutyrate and 50% 3-hydroxyvalerate was expected [17]. However, less energy is needed to metabolize propionic acid than acetic acid [40], which explains the higher percentage of 3-hydroxyvalerate in the PHA.

The non-purified PHA (Figure 5A) contained a bigger variety of impurities than the purified PHA (Figure 5B). Not much can be said about the absolute concentration of impurities before and after the purification process, as no calibration curves were made for the non-PHA related compounds. However, a reduction of 71.4%, 71.6%, and 63.7% in the areas of the impurities III, IV and V, respectively, was calculated, indicating a significant reduction in the overall concentration of such impurities.

Although the quantity of impurities was reduced after the 1-butanol treatments, hexadecanoic acid, octadecanoic acid and dehydroabietic acid were still found in all purified samples. Hexadecanoic and octadecanoic acids have been reported as storage compounds produced by mixed bacterial cultures [41]. The source of the dehydroabietic acid is unknown.

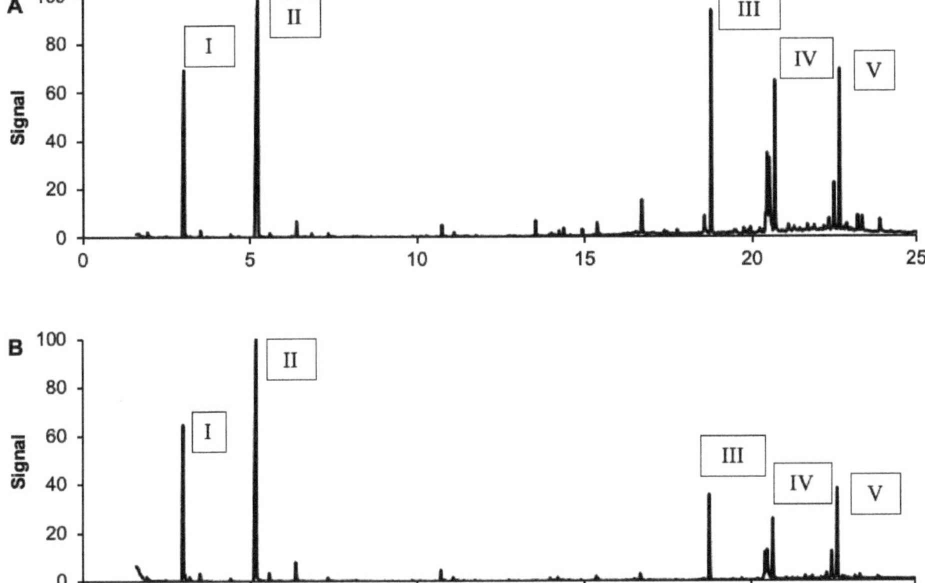

Figure 5. Chromatograms of PHA before (**A**) and after (**B**) purification. The identified compounds are: I—3-hydroxybutyric acid; II—3-hydroxyvaleric acid; III—Hexadecanoic acid; IV—Octadecanoic acid; V—Dehydroabietic acid.

3.5. Applicability

In this work, a mixed microbial culture was used for a PHA accumulation procedure followed by an extraction and purification with different solvents for obtaining a high purity final product. However, the extraction and purification are a two-step process that can be very costly when it comes to an industrial setting. For certain PHA applications where high purity is not a major factor, a single extraction with DMC could be enough for the commercial feasibility of the process. For applications where high purity PHA is required, the purification step can be added, although higher costs should be expected.

4. Conclusions

The extraction of PHA from mixed microbial cultures can be successfully done with dimethyl carbonate via reflux extraction. Overall, a very small variance of PHA yield was observed for different extraction times or biomass to solvent ratios. A ratio of 0.05 g ml^{-1} is considered to be ideal, as the use of higher amounts of biomass lead to practical difficulties. Although higher extraction yields can be obtained with chloroform or dichloromethane as solvents, that also leads to a decrease in the purity of PHA and a less sustainable extraction.

A purification of the extracted PHA with 1-butanol resulted in an increase in purity from 91.2 ± 0.1% to 98.0 ± 0.1%. Although the total purity is approximately the same for different purification times and PHA to solvent ratios, a 0.01 g mL^{-1} ratio for 0.5 h of purification time led to a whiter PHA.

Dimethyl carbonate is a great alternative to conventional hazardous solvents in the extraction process of PHA and 1-butanol can be used to increase the purity of PHA if necessary, leading to a more commercially attractive product, although this could lead to higher production costs. In terms of a circular economy, the whole process opens new environmentally friendly possibilities for the bioplastic industry.

Author Contributions: Conceptualization, G.A.d.S.R., M.H.A.M., G.L.F. and I.L.; Data curation, G.A.d.S.R. and G.L.F.; Formal analysis, G.A.d.S.R., M.H.A.M. and G.L.F.; Funding acquisition, J.H.d.B.; Investigation, G.A.d.S.R. and G.L.F.; Methodology, G.A.d.S.R., M.H.A.M. and G.L.F.; Project administration, J.H.d.B.; Supervision, M.H.A.M. and I.L.; Writing—original draft, G.A.d.S.R. and G.L.F.; Writing—review and editing, G.A.d.S.R., M.H.A.M., I.L. and J.H.d.B. All authors have read and agreed to the published version of the manuscript.

Funding: This work was partially funded by Interreg North-West Europe (European Regional Development Fund) as a part of the WOW! Project (Wider business Opportunities for raw materials from Wastewater) [grant NWE619].

Acknowledgments: The authors would like to thank the technicians at Avans University of Applied Sciences for their technical assistance in this research and the Living Lab Biobased Brazil for connecting the authors. We are also grateful for the editors and the reviewers for their constructive comments which helped to improve the quality of this publication.

Conflicts of Interest: The authors declare no conflict of interest.

Abbreviations

3HB	3-hydroxybutyrate
3HV	3-hydroxyvalerate
COD	Chemical oxygen demand
DMC	Dimethyl carbonate
DO	Dissolved oxygen
GC-MS	Gas chromatography—mass spectrometry
MMC	Mixed Microbial Culture
PHAs	Polyhydroxyalkanoates
PHBV	Poly(3-hydroxybutyrate-co-3-hydroxyvalerate)
TGA	Thermal gravimetric analysis
VFAs	Volatile fatty acids
VSS	Volatile suspended solids

References

1. Laycock, B.; Halley, P.; Pratt, S.; Werker, A.; Lant, P. The chemomechanical properties of microbial polyhydroxyalkanoates. *Prog. Polym. Sci.* **2014**, *39*, 397–442. [CrossRef]
2. Shrivastav, A.; Kim, H.; Kim, Y. Advances in the Applications of Polyhydroxyalkanoate Nanoparticles for Novel Drug Delivery System. *Biomed Res. Int.* **2013**, *2013*, 1–12. [CrossRef] [PubMed]
3. Chen, G.-Q. Plastics from Bacteria: Natural Functions and Applications. *Microbiol. Monogr.* **2009**, *14*, 121–132.
4. Elain, A.; Le Grand, A.; Corre, Y.-M.; Le Fellic, M.; Hachet, N.; Le Tilly, V.; Loulergue, P.; Audic, J.-L.; Bruzaud, S. Valorisation of local agro-industrial processing waters as growth media for polyhydroxyalkanoates (PHA) production. *Ind. Crops Prod.* **2016**, *80*, 1–5. [CrossRef]
5. Koller, M.; Marsalek, L.; de Sousa Dias, M.M.; Braunegg, G. Producing microbial polyhydroxyalkanoate (PHA) biopolyesters in a sustainable manner. *New Biotechnol.* **2017**, *37*, 24–38. [CrossRef]
6. Tamis, J.; Luzkov, K.; Jiang, Y.; van Loosdrecht, M.C.M.; Kleerebezem, R. Enrichment of *Plasticicumulans acidivorans* at pilot-scale for PHA production on industrial wastewater. *J. Biotechnol.* **2014**, *192*, 161–169. [CrossRef]
7. Nikodinovic-Runic, J.; Guzik, M.; Kenny, S.T.; Babu, R.; Werker, A.; Connor, K.E. Carbon-Rich Wastes as Feedstocks for Biodegradable Polymer (Polyhydroxyalkanoate) Production Using Bacteria. *Adv. Appl. Microbiol.* **2013**, *84*, 139–200. [PubMed]
8. Werker, A.G.; Johansson, P.S.T.; Magnusson, P.O.G.; Maurer, F.H.J.; Jannasch, P. Method for Recovery of Stabilized Polyhydroxyalkanoates from Biomass that Has been Used to Treat Organic Waste. US2013/0203954A1, 2013. Available online: https://patents.google.com/patent/US20130203954A1/fr (accessed on 2 March 2020).
9. Bengtsson, S.; Werker, A.; Visser, C.; Korving, L. *PHARIO—Stepping Stone for a Sustainable Value Chain for PHA Bioplastic Using Municipal Activated Sludge*; Stowa: Amersfoort, The Netherlands, 2017; p. 15.

10. Oliveira, C.S.S.; Silva, C.E.; Carvalho, G.; Reis, M.A.M. Strategies for efficiently selecting PHA producing mixed microbial cultures using complex feedstocks: Feast and famine regime and uncoupled carbon and nitrogen availabilities. *New Biotechnol.* **2017**, *37*, 69–79. [CrossRef] [PubMed]

11. Carvalho, G.; Oehmen, A.; Albuquerque, M.G.E.; Reis, M.A.M. The relationship between mixed microbial culture composition and PHA production performance from fermented molasses. *New Biotechnol.* **2014**, *31*, 257–263. [CrossRef]

12. Valentino, F.; Karabegovic, L.; Majone, M.; Morgan-Sagatsume, F.; Werker, A. Polyhydroxyalkanoate (PHA) storage within a mixed-culture biomass with simultaneous growth as a function of accumulation substrate nitrogen and phosphorus levels. *Water Res.* **2015**, *77*, 49–63. [CrossRef]

13. Lemos, P.C.; Serafim, L.S.; Reis, M.A.M. Synthesis of polyhydroxyalkanoates from different short-chain fatty acids by mixed cultures submitted to aerobic dynamic feeding. *J. Biotechnol.* **2006**, *122*, 226–238. [CrossRef] [PubMed]

14. Bengtsson, S. The utilization of glycogen accumulating organisms for mixed culture production of polyhydroxyalkanoates. *Biotechnol. Bioeng.* **2009**, *104*, 698–708. [CrossRef] [PubMed]

15. Dionisi, D.; Levantesi, C.; Renzi, V.; Tandoi, V.; Majone, M. PHA storage from several substrates by different morphological types in an anoxic/aerobic SBR. *Water Sci. Technol.* **2002**, *46*, 337–344. [CrossRef]

16. Kessler, B.; Witholt, B. Factors involved in the regulatory network of polyhydroxyalkanoate metabolism. *J. Biotechnol.* **2001**, *86*, 97–104. [CrossRef]

17. Wang, J.; Yue, Z.-B.; Sheng, G.-P.; Yu, H.-Q. Kinetic analysis on the production of polyhydroxyalkanoates from volatile fatty acids by *Cupriavidus necator* with a consideration of substrate inhibition, cell growth, maintenance, and product formation. *Biochem. Eng. J.* **2010**, *49*, 422–428. [CrossRef]

18. Pardelha, F.; Albuquerque, M.G.E.; Reis, M.A.M.; Dias, J.M.L.; Oliveira, R. Flux balance analysis of mixed microbial cultures: Application to the production of polyhydroxyalkanoates from complex mixtures of volatile fatty acids. *J. Biotechnol.* **2012**, *162*, 336–345. [CrossRef] [PubMed]

19. Albuquerque, M.G.E.; Martino, V.; Pollet, E.; Avérous, L.; Reis, M.A.M. Mixed culture polyhydroxyalkanoate (PHA) production from volatile fatty acid (VFA)-rich streams: Effect of substrate composition and feeding regime on PHA productivity, composition and properties. *J. Biotechnol.* **2011**, *151*, 66–76. [CrossRef]

20. Hao, J.; Wang, H.; Wang, X. Selecting optimal feast-to-famine ratio for a new polyhydroxyalkanoate (PHA) production system fed by valerate-dominant sludge hydrolysate. *Appl. Microbiol. Biotechnol.* **2018**, *102*, 3133–3143. [CrossRef]

21. Chen, G.Q.; Zhang, G.; Park, S.J.; Lee, S.Y. Industrial scale production of poly(3-hydroxybutyrate-co-3-hydroxyhexanoate). *Appl. Microbiol. Biotechnol.* **2001**, *57*, 50–55.

22. Manangan, T.; Shawaphun, S. Quantitative extraction and determination of polyhydroxyalkanoate accumulated in *Alcaligenes latus* dry cells. *Scienceasia* **2010**, *36*, 199–203. [CrossRef]

23. Riedel, S.L.; Brigham, C.J.; Budde, C.F.; Bader, J.; Rha, C.K.; Stahl, U.; Sinskey, A.J. Recovery of poly(3-hydroxybutyrate-co-3-hydroxyhexanoate) from *Ralstonia eutrophacultures* with non-halogenated solvents. *Biotechnol. Bioeng.* **2012**, *110*, 461–470. [CrossRef] [PubMed]

24. Jiang, X.; Ramsay, J.A.; Ramsay, B.A. Acetone extraction of mcl-PHA from *Pseudomonas putida* KT2440. *J. Microbiol. Methods* **2006**, *67*, 212–219. [CrossRef] [PubMed]

25. Fernández-Dacosta, C.; Posada, J.A.; Kleerebezem, R.; Cuellar, M.C.; Ramirez, A. Microbial community-based polyhydroxyalkanoates (PHAs) production from wastewater: Techno-economic analysis and ex-ante environmental assessment. *Bioresour. Technol.* **2015**, *185*, 368–377. [CrossRef] [PubMed]

26. Fei, T.; Cazeneuve, S.; Wen, Z.; Wu, L.; Wang, T. Effective Recovery of Poly-b-Hydroxybutyrate (PHB) Biopolymer from *Cupriavidus necator* Using a Novel and Environmentally Friendly Solvent System. *Biotechnol. Prog.* **2016**, *32*, 678–685. [CrossRef]

27. Samorì, C.; Abbondanzi, F.; Galletti, P.; Giorgini, L.; Mazzocchetti, L.; Torri, C.; Tagliavini, E. Extraction of polyhydroxyalkanoates from mixed microbial cultures: Impact on polymer quality and recovery. *Bioresour. Technol.* **2015**, *189*, 195–202. [CrossRef]

28. Samorì, C.; Basaglia, M.; Casella, S.; Favaro, L.; Galletti, P.; Giorgini, L.; Marchi, D.; Mazzocchetti, L.; Torri, C.; Tagliavini, E. Dimethyl carbonate and Switchable Anionic Surfactants: Two effective tools for the extraction of polyhydroxyalkanoates from microbial biomass. *Green Chem.* **2015**, *17*, 1047–1056. [CrossRef]

29. National Center for Biotechnology Information. PubChem Database. Dimethyl Carbonate, CID:12021. Available online: https://pubchem.ncbi.nlm.nih.gov/compound/Dimethyl-carbonate (accessed on 2 March 2020).

30. Hansch, C.; Leo, A.; Hoekman, D. *Exploring QSAR—Hydrophobic, Electronic and Steric Constants*; American Chemical Society: Washington, DC, USA, 1995; pp. 3–14.

31. Werker, A.G.; Johansson, P.S.T.; Magnusson, P.O.G. Process for the Extraction of Polyhydroxyalkanoates from Biomass. US2015/0368393A1, 2015. Available online: https://patents.google.com/patent/US20150368393/en17 (accessed on 2 March 2020).

32. Carrier, M.; Loppinet-Serani, A.; Denux, D.; Lasnier, J.-M.; Ham-Pichavant, F.; Cansell, F.; Aymonier, C. Thermogravimetric analysis as a new method to determine the lignocellulosic composition of biomass. *Biomass Bioenergy* **2011**, *35*, 298–307. [CrossRef]

33. Lo, C.-W.; Wu, H.-S.; Wei, Y.-H. Optimizing acidic methanolysis of poly(3-hydroxyalkanoates) in gas chromatography analysis. *Asia Pac. J. Chem. Eng.* **2009**, *4*, 487–494. [CrossRef]

34. Hahn, S.K.; Chang, Y.K. A thermogravimetric analysis for poly(3-hydroxybutyrate) quantification. *Biotechnol. Tech.* **1995**, *9*, 873–878. [CrossRef]

35. Bengtsson, S.; Pisco, A.R.; Johansson, P.; Lemos, P.C.; Reis, M.A.M. Molecular weight and thermal properties of polyhydroxyalkanoates produced from fermented sugar molasses by open mixed cultures. *J. Biotechnol.* **2010**, *147*, 172–179. [CrossRef]

36. Carrasco, F.; Dionisi, D.; Beccari, M.; Majone, M.; Martinelli, A. Thermal stability of polyhydroxyalkanotaes. *J. Appl. Polym. Sci.* **2006**, *100*, 2111–2121. [CrossRef]

37. Keenan, T.M.; Tanenbaum, S.W.; Stipanovic, A.J.; Nakas, J.P. Production and characterization of poly-beta-hydroxyalkanoate copolymers from *Burkholderia cepacia* utilizing xylose and levulinic acid. *Biotechnol. Prog.* **2004**, *20*, 1697–1704. [CrossRef] [PubMed]

38. Jiang, Y.; Chen, Y. The effects of the ratio of propionate to acetate on the transformation and composition of polyhydroxyalkanoates with enriched cultures of glycogen-accumulating organisms. *Environ. Technol.* **2009**, *30*, 241–249. [CrossRef] [PubMed]

39. Dai, Y.; Lambert, L.; Yuan, Z.; Keller, J. Characterisation of polyhydroxyalkanoate copolymers with controllable four-monomer composition. *J. Biotechnol.* **2008**, *134*, 137–145. [CrossRef] [PubMed]

40. Montano-Herrera, L.; Laycock, B.; Werker, A.; Pratt, S. The Evolution of Polymer Composition during PHA Accumulation: The Significance of Reducing Equivalents. *Bioengineering* **2017**, *4*, 20. [CrossRef]

41. Castro, A.R.; Silva, P.T.S.; Castro, P.J.G.; Alves, E.; Domingues, M.R.M.; Pereira, M.A. Tuning culturing conditions towards the production of neutral lipids from lubricant-based wastewater in open mixed bacterial communities. *Water Res.* **2018**, *144*, 532–542. [CrossRef]

Article

Quantitative PCR Detection of Enteric Viruses in Wastewater and Environmental Water Sources by the Lisbon Municipality: A Case Study

Pedro Teixeira [1,2,3,*], Sílvia Costa [1], Bárbara Brown [1], Susana Silva [4], Raquel Rodrigues [3] and Elisabete Valério [3]

[1] Câmara Municipal de Lisboa, Direcção Municipal do Ambiente, Estrutura Verde, Clima e Energia, Laboratório de Bromatologia e Águas, Avenida Cidade do Porto S/N-1700-111 Lisboa, Portugal; silvia.a.costa@cm-lisboa.pt (S.C.); barbara.brown@cm-lisboa.pt (B.B.)

[2] Centro de Estudos do Ambiente e do Mar (CESAM), Faculdade de Ciências da Universidade de Lisboa, Campo Grande, 1749-016 Lisboa, Portugal

[3] Departamento de Saúde Ambiental, Instituto Nacional de Saúde Doutor Ricardo Jorge, Avenida Padre Cruz, 1649-016 Lisboa, Portugal; raquel.rodrigues@insa.min-saude.pt (R.R.); elisabete.valerio@insa.min-saude.pt (E.V.)

[4] Departamento de Epidemiologia, Instituto Nacional de Saúde Doutor Ricardo Jorge, Avenida Padre Cruz, 1649-016 Lisboa, Portugal; susana.pereira@insa.min-saude.pt

[*] Correspondence: pedro.teixeira@cm-lisboa.pt

Received: 10 January 2020; Accepted: 13 February 2020; Published: 15 February 2020

Abstract: Current regulations and legislation require critical revision to determine safety for alternative water sources and water reuse as part of the solution to global water crisis. In order to fulfill those demands, Lisbon municipality decided to start water reuse as part of a sustainable hydric resources management, and there was a need to confirm safety and safeguard for public health for its use in this context. For this purpose, a study was designed that included a total of 88 samples collected from drinking, superficial, underground water, and wastewater at three different treatment stages. Quantitative Polimerase Chain Reaction (PCR) detection (qPCR) of enteric viruses Norovirus (NoV) genogroups I (GI) and II (GII) and Hepatitis A (HepA) was performed, and also FIB (*E. coli*, enterococci and fecal coliforms) concentrations were assessed. HepA virus was only detected in one untreated influent sample, whereas NoV GI/ NoV GI were detected in untreated wastewater (100/100%), secondary treated effluent (47/73%), and tertiary treated effluent (33/20%). Our study proposes that NoV GI and GII should be further studied to provide the support that they may be suitable indicators for water quality monitoring targeting wastewater treatment efficiency, regardless of the level of treatment.

Keywords: norovirus; water reuse; water quality

---

1. Introduction

Monitoring every pathogenic microorganism potentially present in water, namely viruses, bacteria, protozoa, or fungi, is unrealistic given the number of resources necessary for that purpose. As an alternative, microbiological water quality assessment has been focused essentially on detecting fecal indicator bacteria (FIB), namely *Escherichia coli* (*E. coli*) and *Enterococcus* spp. [1–6]. FIB are used as surrogates for enteric pathogens in particular for monitoring fecal contamination in environmental waters, relying on the principle that FIB existence is concomitant with pathogen presence, as demonstrated in several studies [7,8]. However, there are also numerous studies showing that pathogens do not correlate significantly with FIB [9–18]. Since FIB, like *E. coli* and *Enterococcus* spp. are shed in most animal feces [4,19,20], the lack of suitability between FIB levels and human health outcomes may be related to the FIB source. Amplifying the disconnection between FIB and pathogens is the ability for FIB not only to persist but to grow in environmental habitats like terrestrial soils, aquatic sediments, and aquatic vegetation [16,21–26]. Further studies are thus essential to assess the suitability of FIB as sole indicators not only in environmental waters but in an area that has been gaining more importance worldwide: treated wastewater use, as demand for water reuse, is increasing worldwide - whether by necessity in developing countries or by environmental objectives in developed countries.

An efficient and sustainable hydric resources management allows non-potable uses for treated wastewater, such as irrigation, industrial processes, firefighting, recreational, or municipal services. Besides the existence of heavy metals, chemicals, hormones and endocrine disruptors in wastewater, it is still necessary to deal with the expected presence of resistant pathogenic microorganisms, many of which are not tested or included in current standards or legislation for water quality assessment. These microorganisms include viruses, bacteria, protozoa, and helminths, responsible for a significant number of potentially dangerous pathologies. There are some reports showing that wastewater treatment processes do not completely remove enteric viruses [27–29], even from effluents with adequate chlorine concentrations or UV treatment [30]. With an increase in reclaimed water use, the potential health impacts resultant from microbial contamination need to be further explored, as outbreaks of viral infectious diseases have been linked to insufficient treatment [31–33]. Several viruses, including Norovirus (NoV) and Hepatitis A (HepA), are listed in United States Environmental Protection Agency (USEPA's) drinking water Contaminant Candidate List (4-CCL 4), heightening waterborne viruses as a research priority [34]. According to a recent review by Teixeira et al. ([35], in press), numerous studies have focused on detecting enteric viruses in water samples, including treated and untreated wastewater, as well as environmental waters. Data on NoV and HepA viruses' concentrations, however, particularly in tertiary-treated wastewater, are scarce ([35], in press). Therefore, further studies are necessary to evaluate the adequate indicators for water reuse quality evaluation and support their application for regulatory purposes.

Globally, NoV is responsible for nearly 20% of all acute gastroenteritis cases [36], with 677 million cases per year and over 213,000 deaths [37]. In risk groups comprising children, elderly, or immunocompromised individuals, morbidity, and mortality rates of NoV infection are significant [38–42]. NoV exposure, and possible outbreaks, have been reported in schools, hospitals, cruise ships, nursing homes, swimming pools, and restaurants [43–47]. Transmission of the virus primarily occurs via fecal-oral contamination route, direct contact with an infected individual, and contaminated water or food consumption [47–51]. Contact with the virus may occur through drinking [52], recreational [44], or irrigation water [53], leading to waterborne outbreaks [54–56]. The vast contaminated aqueous sources linked to NoV indicate a ubiquitous distribution of the virus [57]. HepA virus transmission occurs mainly by the oral-fecal route (about 95%), and ingestion of contaminated water and food [58–62]. HepA virus has been detected in drinking water, surface water, groundwater, treated, and untreated wastewater [63–67]. With global distribution, HepA is a major causal agent of acute viral hepatitis, with approximately 1.4 million cases reported annually globally [60,68], and a high endemicity of hepatitis A in regions with low sanitation [60,69,70].

Taking benefit from an existing standardized method for detecting NoV and HepA viruses in food and bottled water using real-time quantitative PCR detection (RT-qPCR) (ISO/TS, 15216-1:2017), the aim of our study was to develop a method based on this standardized method and assess the suitability of these enteric viruses as water quality indicators for water use and reuse. Moreover, the method developed aimed to be suitable for several water matrices—drinking, underground, superficial, treated, and untreated wastewater, thus enabling its use in routine water quality testing for detecting and quantifying NoV GI/GII and HepA viruses, and allowing managing entities to evaluate more accurately potential public health risks.

2. Materials and Methods

2.1. Sampling

Fieldwork was carried out from February 2018 to December 2018, with a total of 88 samples collected. Sampling included: 1) underground water samples collected at two different sites in Lisbon (n = 19); 2) superficial water intended for drinking water production (n = 10); 3) drinking water from Lisbon's supply system (n = 11); 4) wastewater collected at a Wastewater Treatment Plant in the Lisbon district, at three different stages – untreated influent (n = 15), effluent with secondary treatment (n = 15) and effluent with tertiary, sand filtration and UV treatment (n = 15). Additionally, blank assays were performed with distilled, sterilized water samples (n = 3). All samples were collected in sterile polyethylene containers and stored at 4 °C for less than 2 h until chemical and microbiological determinations were initiated. 0.5 mL of a 10% dechlorination agent $Na_2S_2O_3$ was added to containers before sampling in order to neutralize possible existing residual chlorine from drinking water samples. Water samples for enteric viruses' determinations were collected in sterile glass containers. Enteric viruses were determined in different volumes according to the water source (25–5000 mL). Microbiological indicators were determined in 100 mL, and chemical indicators were determined according to standard procedures, from a total of 1000 mL [71]. The time lapse between sample collection and laboratory processing did not exceed 24 h.

2.2. Detection and Quantification of Enteric Viruses

The procedure established in this study was based on the international standard method for the determination of viruses in foods ISO/TS 15216-1:2017 [72]. Initially, 10 µL of Mengo virus strain vMC0 (ceeramTOOLS®, Biomérieux, France) were added to each sample, to be used as an internal process control to assure the RNA extraction efficiency. Samples with extraction efficiency, ≥1% were considered valid, as established in Norm ISO/TS 15216-1:2017 [72]. Viral particles were captured through filtration with a 0.45 µm pore size (47 mm diameter) positively charged nylon membrane (Amersham Hybond N+, GE Healthcare Life Sciences, UK), from 25 mL (untreated wastewater), 1000 mL (blank assay, secondary, tertiary treated wastewater and superficial water) and 5000 mL (drinking and underground water) samples. It should be noted that there are existing alternative protocols for virus collection [73], which stipulate significantly high volumes for filtration and virus collection in surface and groundwater—300 L and 1500 L, respectively. For practical reasons, namely to avoid filter clogging and the possible inexistence of such water volumes in some groundwater sources, and more importantly, aiming to establishing a reasonably achievable method for detecting and quantification enteric viruses in diverse water samples for laboratories that perform the routine water monitoring, in this study different volumes were tested and filtered according to the sample matrix. Filters were then transferred into a sterile tube, and 4 mL of tris/glycine/beef extract (TGBE) buffer were added and shaken at approximately 50 oscillations min^{-1} for 20 ± 5 min. An additional 4 mL of TGBE buffer was added, and the eluates pooled into a single clean tube. The pH was adjusted to 7.0 ± 0.5 with HCl (≥5 mol/L). Subsequently, samples were concentrated using Amicon® Ultra-15 Centrifugal Filter Devices with a 100 kDa molecular weight cut-off (Merck Millipore, Darmstadt, Germany), through centrifugation 4000 g, at 4 °C, for 15 min. For RNA isolation, 500 µL TRIzol® reagent (Thermo Fisher Scientific, Waltham, MA, USA) was added to the samples and mixed for 5 min, 30 °C at 350 rpm in a thermomixer (Eppendorf, Germany). Afterward, 200 µL of chloroform were added and samples were mixed in a thermomixer (Eppendorf, Germany) for 3 min, 30 °C at 350 rpm. Centrifugation was then performed for 15 min, at 4 °C and 12,000 g. RNA was extracted from 140 µL final volume (approx.) of the concentrated sample (aqueous phase) using a QIAamp Viral RNA Mini kit (QIAGEN, Hilden, Germany), according to the manufacturer's instructions. Samples were stored at −80 °C until analysis. Quantitative Real-Time PCR (RT-qPCR) was performed (Applied Biosystems AB7500 qPCR) for the specific detection of NoV GI, NoV GII, and HepA viruses. An initial screening for the selected viruses was performed, in addition to confirmation for the absence of inhibitors with

10-fold dilutions. RT-qPCR was performed for NoV GI, NoV GII, HepA, and Mengo virus detection and quantification, with the use of commercial kits (ceeramTOOLS®, Biomérieux, France), according to the manufacturer's specifications.

2.3. Microbiological Analysis

Detection of total coliforms, E. coli, enterococci, and fecal coliforms was performed through the use of Colilert and Enterolert with Quanti-Tray (IDDEX Laboratories, Westbrook, ME, USA). Samples were processed according to manufacturer's instructions. For heterotrophic plate count at 22 °C and 37 °C, agar inclusion technique was performed using 1 mL aliquots of the water sample (after the necessary dilutions were performed) and adding 15 mL of Yeast Extract agar (VWR Chemicals, Radnor, PA, USA), and incubated at 36 °C for 44 h and 22 °C for 68 h. After the incubation period at each temperature, all colonies were quantified for each case.

2.4. Physical and Chemical Assessment

Temperature and pH were both determined (Thermo Scientific™ Orion™3-Star Benchtop pH Meter, Thermo Fisher Scientific, Waltham, MA, USA) according to standard methods, as well as free chlorine and total organic carbon (TOC) [71]. Conductivity was measured according to Standard Guideline NP EN 27888:1996 (MeterLab CDM 210) and Mohr's Method was performed to assess chlorides.

2.5. Statistical Analysis

A descriptive analysis was made for numeric and categorical variables. For comparing different measures among the three wastewater treatment phases, a Kruskal–Wallis non-parametric or Fisher's Exact Test was used. When differences were found, multiple comparisons with Bonferroni correction was also performed. Correlation in each wastewater treatment phase was measured with the Pearson correlation coefficient. Statistical tests were performed bilaterally at a significance level of 5%. The statistical analysis of the data was performed using statistical software R, version 3.4.3. [74].

3. Results

3.1. Environmental and Drinking Waters

A statistical summary for virus, microbiological, physical and chemical results (median) is presented in Table 1, concerning blank assays (BA), superficial water intended for drinking water production (DW1), drinking water from Lisbon's public supply system (DW2) and underground water samples collected at two different sites in Lisbon (GW1 and GW2). None of the enteric viruses targeted in this study—NoV GI, NoV GII, and HepA—were detected in samples from superficial water intended for drinking water production (n = 10), drinking water from Lisbon's supply system (n = 11) and underground water samples (n = 19).

Microbiological, physical, and chemical results for drinking water samples (n = 11) were all in compliance with national legislation—Law Decree No. 306/2007, 27th August, and Law Decree No. 236/1998, 1st August, with no microbial contamination detected. Environmental - superficial and underground bodies of water - presented similar and reduced levels of fecal contamination, with no significant differences between them (Table 1). Apart from conductivity and TOC, which presented lower and higher values, respectively, in the superficial water samples. The physical and chemical results were similar between the superficial and underground water samples.

Table 1. Microbiological, physical, and chemical results (median) for blank assays (BA), superficial water intended for drinking water production (DW1), drinking water from Lisbon's supply system (DW2), and underground water samples collected at two different sites in Lisbon (GW1 and GW2). MNP—Most Probable Number, CFU—Colony Forming Units, GC—Genomic Copies. ND—Not detected. n = total analyzed samples. LD (limit of detection) = 1 MPN/100 mL.

	BA	DW1	DW2	GW1	GW2
	(n = 3)	(n = 10)	(n = 11)	(n = 8)	(n = 11)
Total Coliforms (MPN/100 mL)	<LD	7.58×10^2	<LD	5.6×10^1	2.01×10^2
E. coli (MPN/100 mL)	<LD	2	<LD	3	<LD
Fecal Coliforms (MPN/100 mL)	<LD	1	<LD	2	1
Enterococcus (MPN/100 mL)	<LD 0	4	<LD	5	2.6×10^1
HPC37 °C (CFU/mL)	ND	8.4×10^1	ND	1.4×10^1	1.57×10^2
HPC22 °C (CFU/mL)	ND	1.86×10^2	ND	8.9×10^1	3.01×10^2
TOC (mg/L)	0.0	3.9	0.0	0.0	0.0
Chlorides (mg Cl$^-$/L)	1	7	14	30	69
Free chlorine (mg Cl$_2$/L)	0.06	0.06	0.60	0.08	0.07
Conductivity (µS/cm at 20 °C)	2	71	173	552	827
pH (25°)	6.0	7.9	7.9	7.8	7.1
NoV GI (GC/100 mL)	ND	ND	ND	ND	ND
NoV GII (GC/100 mL)	ND	ND	ND	ND	ND
HepA (GC/100 mL)	ND	ND	ND	ND	ND

3.2. Wastewater

Untreated wastewater influent (WW1), effluent with secondary treatment (WW2) and effluent with tertiary treatment (WW3) statistical results summary for virus, microbiological, physical and chemical results (median) is presented in Table 2. HepA virus was only detected in one untreated influent sample, with a 3.99×10^6 gc/100 mL concentration. NoV GI was detected in all wastewater treatment stages - untreated wastewater influent (n = 15), effluent with secondary treatment (n = 7) and effluent with tertiary treatment (n = 5). Similar results were obtained for NoV GII with the following positive samples: untreated wastewater influent (n = 15), effluent with secondary treatment (n = 11) and effluent with tertiary treatment (n = 3). It can be noted that NoV GI median concentration reductions are significant between stages; a 6.67×10^7 gc/100 mL initial concentration is reduced to 8.34×10^6 gc/100 mL (WW2) and to 8.81×10^5 gc/100 mL final concentration. NoV GII initial 4.86×10^7 gc/100 mL median concentration is also significantly reduced with secondary treatment to 2.29×10^6 gc/100 mL (WW2). With UV and sand filtration, the effluent's concentration is also diminished (9.69×10^5 gc/100 mL), although with no observed statistical significance.

Significant differences were found between all treatment wastewater phases for FIB and HPC at 37 °C and 22 °C concentrations. Notably, we can observe a significant FIB decay not only with secondary treatment but also with tertiary sand filtration and UV treatment—after which FIB have been eliminated upon tertiary treatment (results < limit of detection). According to current Portuguese national legislations—Law Decree No. 119/2019, 21st August—these effluents (with tertiary treatment) can be adequate for non-potable uses including irrigation, industrial uses, firefighting or street cleaning. Excluding chlorides, with no significant variations between wastewater treatments, the majority of the physical and chemical results, display a statistically significant reduction with secondary treatment - activated sludge.

Table 2. Microbiological, physical, and chemical results (median) for untreated wastewater influent (WW1), effluent with secondary treatment (WW2) and effluent with tertiary treatment (WW3). * Kruskall–Wallis non parametric test. ** Multiple comparisons tests with Bonferroni correction. a—Statistically significant difference between untreated influent and secondary treated effluent. b—Statistically significant difference between untreated influent and tertiary treated effluent. c—Statistically significant difference between secondary and tertiary treated effluent. MNP—Most Probable Number, CFU—Colony Forming Units, GC—Genomic Copies. ND—Not Detected. n = total analyzed samples. LD (limit of detection) = 1 MPN/100 mL.

	WW1	WW2	WW3	P-Value *	Pairwise Test **
	(n = 15)	(n = 15)	(n = 15)		
Total Coliforms (MPN/100 mL)	7.70×10^7	9.80×10^4	<LD	<0.001	a, b, c
E. coli (MPN/100 mL)	1.66×10^7	2.14×10^4	<LD	<0.001	a, b, c
Fecal Coliforms (MPN/100 mL)	1.19×10^7	1.58×10^4	<LD	<0.001	a, b, c
Enterococcus (MPN/100 mL)	3.99×10^6	8.80×10^3	<LD	<0.001	a, b, c
HPC37 °C (CFU/mL)	3.10×10^6	1.50×10^4	4	<0.001	a, b, c
HPC22 °C (CFU/mL)	4.20×10^6	2.04×10^4	2	<0.001	a, b, c
TOC (mg/L)	169.0	7.90	7.70	<0.001	a, b
Chlorides (mg Cl$^-$/L)	173	136	129	0.250	-
Free chlorine (mg Cl$_2$/L)	0.06	0.10	0.09	<0.001	a, b
Conductivity (µS/cm at 20 °C)	1262	839	860	<0.001	a, b
pH (25°)	8.0	7.0	7.5	<0.001	a, b, c
NoV GI (GC/100 mL)	6.67×10^7	8.34×10^6	8.81×10^5	<0.001	a, b, c
NoV GII (GC/100 mL)	4.86×10^7	2.29×10^6	9.69×10^5	<0.001	a, b
HepA (GC/100 mL)	3.99×10^6	ND	ND	-	-

Correlations between the microbiological parameters analyzed in each wastewater treatment phase (WW1, WW2 and WW3) were measured (Figure 1A–C). A positive correlation was observed between NoV GI and NoV GII in untreated influent ($\rho = 0.54$), and between NoV GI and fecal coliforms ($\rho = 0.61$). One of the highest correlations on this phase was found between Total Coliforms and Enterococcus ($\rho = 0.68$). Several positive correlations were detected for secondary treated effluent samples in microbiological parameters but none significant concerning NoV GI/GII. In tertiary treated effluents, only one moderate correlation was observed for HPC 22 °C/37 °C ($\rho = 0.60$).

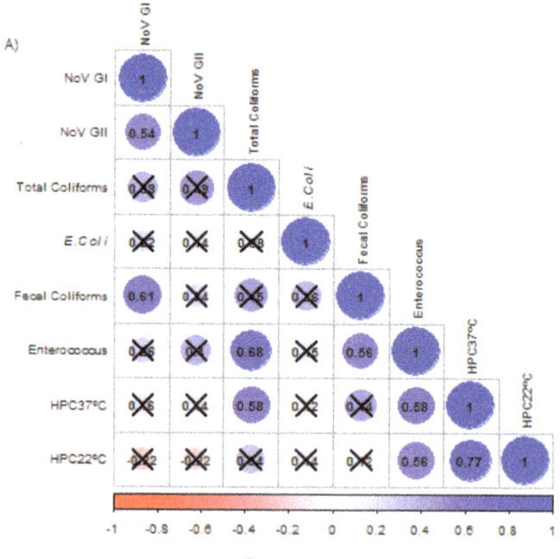

Figure 1. *Cont.*

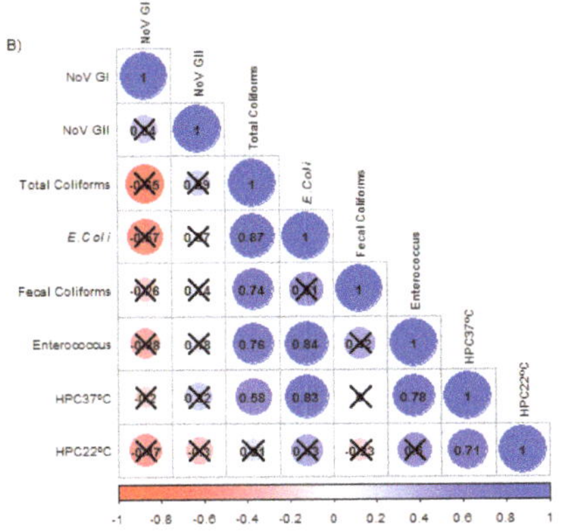

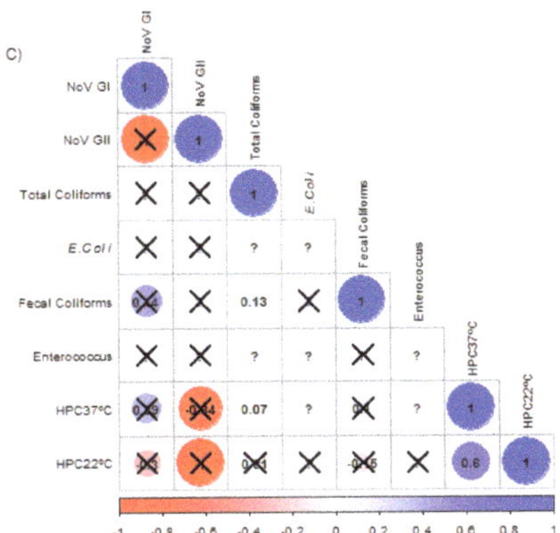

Figure 1. Graphical display of the correlation matrix of the microbiological parameters tested in each wastewater treatment phase (**A**) WW1, (**B**) WW2, and (**C**) WW3. The "?"—means "Not enough finite observations".

4. Discussion and Conclusions

In the case of environmental and drinking waters, enteric viruses NoV GI, NoV GII, and HepA virus were not detected in any of the samples in drinking, superficial, or underground water, suggesting an absence of fecal contamination from a human source in these types of water. However, the presence of total coliforms, *E. coli*, enterococci, and fecal coliforms were detected in samples from underground and superficial samples, and though in reduced concentrations, it suggests a natural and expected fecal contamination from animal origin. Studies showed that NoV, could be detected in particularly high concentrations in feces - up to 10^{11} GC/g [75]. Consequently, the occurrence of NoV in surface waters can be linked to contaminated water sources (i.e., fecal contamination) and with environmental conditions determining the survival of the virus [14,76]. Since these viruses are highly infectious [49] and demonstrate a high resistance to environmental degradation in water [77–79], its presence in surface and groundwater intended for human consumption raises as a potential public health risk, by which frequent monitoring of enteric viruses such as NoV besides FIB, could be desirable and prudent.

On what concerns the treated wastewaters, our results demonstrate that the absence of FIB does not imply the absence of pathogenic microorganisms, namely NoV GI and GII in tertiary treated wastewater (obtained after sand filtration and UV). FIB monitoring and their correlations with enteric viruses have recently been questioned [17,80,81], hence the necessity for more adequate indicators in water quality monitoring, especially for wastewater use. Growing water resources management concerns arise, not only in areas affected by droughts but also importantly in urban areas, like the one targeted in this study—Lisbon, where water reuse represents one of the municipality's commitments as part of a sustainable hydric resources management, with self-evident safeguard for public health. The results here obtained for Nov GI and GII concentrations are in accordance with previous studies targeting untreated, secondary, and tertiary treated wastewater [82–85].

The results of our study support the hypothesis that NoV GI and GII might be suitable indicators for water quality monitoring regarding wastewater treatment efficiency, regardless of the level of the treatment—secondary or tertiary. A correlation between NoV (GI) and FIB (fecal coliforms) is observed in untreated effluent, but in subsequent treatment phases, there was no correlation between the targeted enteric viruses and FIB. While positive correlations observed between FIB and NoV in untreated wastewater could be expected and were indeed observed, the absence of correlations between FIB and secondary and tertiary treatment stages points out to a lack of a link between FIB and the targeted enteric viruses concentrations. Moreover, in several samples with an apparent total FIB elimination with UV and sand filtration treatment, NoV was still detectable. Our results point out to the need of further studies and expansion to different WWTPs, in diverse locations and including a higher number of samples, not only to further reinforce the results of this study and validate our hypothesis but also to expand detection methods to other viral pathogens and evaluate the best and most comprehensive indicators to use for water reuse situations. While cell culture methods have been the gold standard for the detection of infectious viruses, qPCR methods became essential for enteric virus detection in water samples due to shorter detection times, high sensitivity and specificity, and the ability to detect viruses that are not easily or at all culturable [86]. A significant limitation of the method applied in our study is the inability to differentiate infective viruses from non-infective viruses. Nonetheless, viral genomes are present in water samples, particularly in reclaimed wastewater. Considering an estimated 50% infectious dose (ID50) between 18 and 1,015 genomic equivalents [49], public health risks may arise even from non-potable uses like irrigation or street cleaning, as incorrect usage of treated wastewater has caused outbreaks of viral infectious diseases worldwide [31–33]. To this fact, it must enhance while qPCR delivers quantitative data with high accuracy; results obtained by this method should be interpreted prudently because of potential losses during the sample concentration/extraction/purification procedure, which may result in an underestimation in the detection/quantification process [86]. The whole process control used in our study – Mengo virus – allowed evaluating the efficiency for virus recovery in the different stages – from sample concentration and RNA extraction, to RT-qPCR. However, even within the established ISO values (recovery ≥1%), the obtained recovery rates for this control were usually low (<10%), which can underestimate NoV GI and GII concentrations, particularly in reclaimed wastewater. Importantly, this study shows that a procedure based on the international standard method for the determination of viruses in foods ISO/TS 15216-1 [72] could be applied for routine water quality monitoring of NoV and HepA, in different water matrices.

The importance of FIB is undoubtedly as they are responsible for a significant improvement in water quality assessment and safety management for many decades [1]. Nevertheless, current regulations and legislation require critical revision to determine safety in particular for water reuse, as we can observe samples determined compliant for bacterial indicators that are positive for NoV GI and GII presence. Representing a key to the global water crisis, water reuse should also be determined safe considering the different treatments applied, as enteric viruses' infections also represent a worldwide economic concern.

Author Contributions: Conceptualization, E.V. and P.T.; methodology, E.V. and P.T.; validation, E.V. and P.T; formal analysis, P.T., E.V. and S.S.; investigation, P.T., S.C., B.B. and E.V.; resources, P.T., R.R. and E.V.; data curation, P.T. and E.V.; writing—original draft preparation, P.T.; writing—review and editing, E.V., P.T., S.C., S.S., B.B and R.R.; visualization, P.T., S.S. and E.V.; supervision, E.V.; project administration, E.V. and P.T.; funding acquisition, P.T. and E.V. All authors have read and agreed to the published version of the manuscript.

Funding: This research was funded by the Lisbon Municipality (CML) in collaboration with Instituto Nacional de Saúde Doutor Ricardo Jorge (INSA).

Acknowledgments: The authors are grateful to Ângelo Mesquita and José Canêdo for their contributions to the implementation of this project. We would also like to thank Helena Rebelo and João Brandão for all the help provided. Any opinions, findings and conclusions expressed are those of the authors and do not necessarily reflect the views of the Lisbon Municipality.

Conflicts of Interest: The authors declare no conflict of interest.

References

1. Tallon, P.; Magajna, B.; Lofranco, C.; Leung, K.T. Microbial Indicators of Faecal Contamination in Water: A Current Perspective. *Water. Air. Soil Pollut.* **2005**, *166*, 139–166. [CrossRef]
2. José Figueras, M.; Borrego, J.J. New perspectives in monitoring drinking water microbial quality. *Int. J. Environ. Res. Public Health* **2010**, *7*, 4179–4202. [CrossRef] [PubMed]
3. Saxena, G.; Bharagava, R.N.; Kaithwas, G.; Raj, A. Microbial indicators, pathogens and methods for their monitoring in water environment. *J. Water Health* **2015**, *13*, 319–339. [CrossRef] [PubMed]
4. Leclerc, H.; Mossel, D.A.A.; Edberg, S.C.; Struijk, C.B. Advances in the Bacteriology of the Coliform Group: Their Suitability as Markers of Microbial Water Safety. *Annu. Rev. Microbiol.* **2001**, *55*, 201–234. [CrossRef]
5. WHO. Guidelines for safe recreational water environments—Coastal and Fresh waters. *Guidel. Safe Recreat. Water.* **2003**, *1*, 253.
6. EPA. Recreational Water Quality Criteria. Available online: https://www.epa.gov/sites/production/files/2015-10/documents/rwqc2012.pdf (accessed on 13 February 2020).
7. Wade, T.J.; Sams, E.; Brenner, K.P.; Haugland, R.; Chern, E.; Beach, M.; Wymer, L.; Rankin, C.C.; Love, D.; Li, Q.; et al. Rapidly measured indicators of recreational water quality and swimming-associated illness at marine beaches: A prospective cohort study. *Environ. Heal. A Glob. Access Sci. Source* **2010**, *9*, 1–14. [CrossRef]
8. Zmirou, D.; Pena, L.; Ledrans, M.; Letertre, A. Risks associated with the microbiological quality of bodies of fresh and marine water used for recreational purposes: Summary estimates based on published epidemiological studies. *Arch. Environ. Health* **2003**, *58*, 703–711. [CrossRef]
9. Lund, V. Evaluation of E. Coli as an indicator for the presence of Campylobacter jejuni and Yersinia enterocolitica in chlorinated and untreated oligotrophic lake water. *Water Res.* **1996**, *30*, 1528–1534. [CrossRef]
10. Bonadonna, L.; Briancesco, R.; Ottaviani, M.; Veschetti, E. Occurrence of Cryptosporidium oocysts in sewage effluents and correlation with microbial, chemical and physical water variables. *Environ. Monit. Assess.* **2002**, *75*, 241–252. [CrossRef]
11. Lemarchand, K.; Lebaron, P. Occurrence of Salmonella spp. and Cryptosporidium spp. in a French coastal watershed: Relationship with fecal indicators. *FEMS Microbiol. Lett.* **2003**, *218*, 203–209. [CrossRef]
12. Harwood, V.J.; Levine, A.D.; Scott, T.M.; Chivukula, V.; Lukasik, J.; Farrah, S.R.; Rose, J.B. Validity of the indicator organism paradigm for pathogen reduction in reclaimed water and public health protection. *Appl. Environ. Microbiol.* **2005**, *71*, 3163–3170. [CrossRef]
13. Anderson, K.L.; Whitlock, J.E.; Valerie, J.; Harwood, V.J. Persistence and Differential Survival of Fecal Indicator Bacteria in Subtropical Waters and Sediments. *Appl. Environ. Microbiol.* **2005**, *71*, 3041–3048. [CrossRef]
14. Hörman, A.; Rimhanen-finne, R.; Maunula, L.; Von Bonsdorff, C.; Torvela, N. Indicator Organisms in Surface Water in and Indicator Organisms in Surface Water in Southwestern Finland, 2000–2001. *Appl. Environ. Microbiol.* **2004**, *70*, 87–95. [CrossRef]
15. Zhang, Q.; Eichmiller, J.J.; Staley, C.; Sadowsky, M.J.; Ishii, S. Correlations between pathogen concentration and fecal indicator marker genes in beach environments. *Sci. Total Environ.* **2016**, *15*, 826–830. [CrossRef]
16. Bradshaw, J.K.; Snyder, B.J.; Oladeinde, A.; Spidle, D.; Berrang, M.E.; Meinersmann, R.J.; Oakley, B.; Sidle, R.C.; Sullivan, K.; Molina, M. Characterizing relationships among fecal indicator bacteria, microbial source tracking markers, and associated waterborne pathogen occurrence in stream water and sediments in a mixed land use watershed. *Water Res.* **2016**, *101*, 498–509. [CrossRef]

17. McQuaig, S.; Griffith, J.; Harwood, V.J. Association of fecal indicator bacteria with human viruses and microbial source tracking markers at coastal beaches impacted by nonpoint source pollution. *Appl. Environ. Microbiol.* **2012**, *78*, 6423–6432. [CrossRef]
18. Arnold, B.F.; Schiff, K.C.; Griffith, J.F.; Gruber, J.S.; Yau, V.; Wright, C.C.; Wade, T.J.; Burns, S.; Hayes, J.M.; Mcgee, C.; et al. Swimmer illness associated with marine water exposure and water quality indicators: Impact of widely used assumptions. *Epidemiology* **2013**, *24*, 845–853. [CrossRef]
19. Harwood, V.J.; Butler, J.; Parrish, D.; Wagner, V. Isolation of fecal coliform bacteria from the diamondback terrapin (Malaclemys terrapin centrata). *Appl. Environ. Microbiol.* **1999**, *65*, 865–867. [CrossRef]
20. Field, K.G.; Samadpour, M. Fecal source tracking, the indicator paradigm, and managing water quality. *Water Res.* **2007**, *41*, 3517–3538. [CrossRef]
21. Solo-Gabriele, H.M.; Wolfert, M.A.; Desmarais, T.R.; Palmer, C.J. Sources of Escherichia coli in a coastal subtropical environment. *Appl. Environ. Microbiol.* **2000**, *66*, 230–237. [CrossRef]
22. Byappanahalli, M.N.; Shively, D.A.; Nevers, M.B.; Sadowsky, M.J.; Whitman, R.L. Growth and survival of Escherichia coli and enterococci populations in the macro-alga Cladophora (Chlorophyta). *FEMS Microbiol. Ecol.* **2003**, *46*, 203–211. [CrossRef]
23. Jeng, H.C.; England, A.J.; Bradford, H.B. Indicator organisms associated with stormwater suspended particles and estuarine sediment. *J. Environ. Sci. Heal. Part A Toxic/Hazardous Subst. Environ. Eng.* **2005**, *40*, 779–791. [CrossRef] [PubMed]
24. Badgley, B.D.; Thomas, F.I.M.; Harwood, V.J. Quantifying environmental reservoirs of fecal indicator bacteria associated with sediment and submerged aquatic vegetation. *Environ. Microbiol.* **2011**, *13*, 932–942. [CrossRef] [PubMed]
25. Eifan, S.A. Enteric Viruses And Aquatic Environment. *Internet J. Microbiol.* **2013**, *12*, 1–7.
26. He, X.; Wei, Y.; Cheng, L.; Zhang, D.; Wang, Z. Molecular detection of three gastroenteritis viruses in urban surface waters in Beijing and correlation with levels of fecal indicator bacteria. *Environ. Monit. Assess.* **2012**, *184*, 5563–5570. [CrossRef]
27. Blatchley, E.R.; Gong, W.-L.; Alleman, J.E.; Rose, J.B.; Huffman, D.E.; Otaki, M.; Lisle, J.T. Effects of Wastewater Disinfection on Waterborne Bacteria and Viruses. *Water Environ. Res.* **2007**, *79*, 81–92. [CrossRef]
28. El-Senousy, W.M.; Guix, S.; Abid, I.; Pintó, R.M.; Bosch, A. Removal of astrovirus from water and sewage treatment plants, evaluated by a competitive reverse transcription-PCR. *Appl. Environ. Microbiol.* **2007**, *73*, 164–167. [CrossRef]
29. Lazarova, V.; Levine, B.; Sack, J.; Cirelli, G.; Jeffrey, P.; Muntau, H.; Salgot, M.; Brissaud, F. Role of water reuse for enhancing integrated water management in Europe and Mediterranean countries. *Water Sci. Technol.* **2001**, *43*, 25–33. [CrossRef]
30. Sano, D.; Amarasiri, M.; Hata, A.; Watanabe, T.; Katayama, H. Risk management of viral infectious diseases in wastewater reclamation and reuse: Review. *Environ. Int.* **2016**, *91*, 220–229. [CrossRef]
31. Bernard, H.; Faber, M.; Wilking, H.; Haller, S.; Höhle, M.; Schielke, A.; Ducomble, T.; Siffczyk, C.; Merbecks, S.; Fricke, G.; et al. Large multistate outbreak of norovirus gastroenteritis associated with frozen strawberries, Germany, 2012. *Eurosurveillance* **2014**, *19*, 20719. [CrossRef]
32. Okoh, A.I.; Sibanda, T.; Gusha, S.S. Inadequately treated wastewater as a source of human enteric viruses in the environment. *Int. J. Environ. Res. Public Health* **2010**, *7*, 2620–2637. [CrossRef] [PubMed]
33. Sinclair, R.G.; Jones, E.L.; Gerba, C.P. Viruses in recreational water-borne disease outbreaks: A review. *J. Appl. Microbiol.* **2009**, *107*, 1769–1780. [CrossRef] [PubMed]
34. United States Environmental Protection Agency. Contaminant Candidate List (CCL) and Regulatory Determination. Available online: https://www.epa.gov/ccl/microbial-contaminants-ccl-4 (accessed on 30 January 2020).
35. Teixeira, P.; Salvador, D.; Brandão, J.; Ahmed, W.; Sadowsky, M.J.; Valério, E. Environmental and Adaptive Changes Necessitate a Paradigm Shift for Indicators of Fecal Contamination. *Microbiol Spectr.*. in press.
36. Ahmed, S.M.; Hall, A.J.; Robinson, A.E.; Verhoef, L.; Premkumar, P.; Parashar, U.D.; Koopmans, M.; Lopman, B.A. Global prevalence of norovirus in cases of gastroenteritis: A systematic review and meta-analysis. *Lancet Infect. Dis.* **2014**, *14*, 725–730. [CrossRef]
37. Pires, S.M.; Fischer-Walker, C.L.; Lanata, C.F.; Devleesschauwer, B.; Hall, A.J.; Kirk, M.D.; Duarte, A.S.R.; Black, R.E.; Angulo, F.J. Aetiology-specific estimates of the global and regional incidence and mortality of diarrhoeal diseases commonly transmitted through food. *PLoS ONE* **2015**, *10*, e0142927. [CrossRef]

38. Harris, J.P.; Edmunds, W.J.; Pebody, R.; Brown, D.W.; Lopman, B.A. Deaths from norovirus among the elderly, England and Wales. *Emerg. Infect. Dis.* **2008**, *14*, 1546–1552. [CrossRef]

39. Hutson, A.M.; Atmar, R.L.; Estes, M.K. Norovirus disease: Changing epidemiology and host susceptibility factors. *Trends Microbiol.* **2004**, *12*, 279–287. [CrossRef]

40. Payne, D.C.; Vinjé, J.; Szilagyi, P.G.; Edwards, K.M.; Staat, M.A.; Weinberg, G.A.; Hall, C.B.; Chappell, J.; Bernstein, D.I.; Curns, A.T.; et al. Norovirus and medically attended gastroenteritis in U.S. children. *N. Engl. J. Med.* **2013**, *368*, 1121–1130. [CrossRef]

41. Siebenga, J.J.; Vennema, H.; Zheng, D.; Vinjé, J.; Lee, B.E.; Pang, X.; Ho, E.C.M.; Lim, W.; Choudekar, A.; Broor, S.; et al. Norovirus Illness Is a Global Problem: Emergence and Spread of Norovirus GII.4 Variants, 2001–2007. *J. Infect. Dis.* **2009**, *200*, 802–812. [CrossRef]

42. Tian, G.; Jin, M.; Li, H.; Li, Q.; Wang, J.; Duan, Z.J. Clinical characteristics and genetic diversity of noroviruses in adults with acute gastroenteritis in Beijing, China in 2008–2009. *J. Med. Virol.* **2014**, *86*, 1235–1242. [CrossRef]

43. Fankhauser, R.L.; Monroe, S.S.; Noel, J.S.; Humphrey, C.D.; Bresee, J.S.; Parashar, U.D.; Ando, T.; Glass, R.I. Epidemiologic and Molecular Trends of "Norwalk-like Viruses" Associated with Outbreaks of Gastroenteritis in the United States. *J. Infect. Dis.* **2002**, *186*, 1–7. [CrossRef] [PubMed]

44. Podewils, L.J.; Zanardi Blevins, L.; Hagenbuch, M.; Itani, D.; Burns, A.; Otto, C.; Blanton, L.; Adams, S.; Monroe, S.S.; Beach, M.J.; et al. Outbreak of norovirus illness associated with a swimming pool. *Epidemiol. Infect.* **2007**, *135*, 827–833. [CrossRef] [PubMed]

45. White, P.A. Evolution of norovirus. *Clin. Microbiol. Infect.* **2014**, *20*, 741–745. [CrossRef] [PubMed]

46. Xerry, J.; Gallimore, C.I.; Iturriza-Gómara, M.; Allen, D.J.; Gray, J.J. Transmission events within outbreaks of gastroenteritis determined through analysis of nucleotide sequences of the P2 domain of genogroup II noroviruses. *J. Clin. Microbiol.* **2008**, *46*, 947–953. [CrossRef] [PubMed]

47. Widdowson, M.A.; Sulka, A.; Bulens, S.N.; Beard, R.S.; Chaves, S.S.; Hammond, R.; Salehi, E.D.P.; Swanson, E.; Totaro, J.; Woron, R.; et al. Norovirus and foodborne disease, United States, 1991–2000. *Emerg. Infect. Dis.* **2005**, *11*, 95–102. [CrossRef]

48. Kotwal, G.; Cannon, J.L. Environmental persistence and transfer of enteric viruses. *Curr. Opin. Virol.* **2014**, *4*, 37–43. [CrossRef]

49. Teunis, P.F.M.; Moe, C.L.; Liu, P.; Miller, S.E.; Lindesmith, L.; Baric, R.S.; Le Pendu, J.; Calderon, R.L. Norwalk virus: How infectious is it? *J. Med. Virol.* **2008**, *80*, 1468–1476. [CrossRef]

50. Bitler, E.J.; Matthews, J.E.; Dickey, B.W.; Eisenberg, J.N.S.; Leon, J.S. Norovirus outbreaks: A systematic review of commonly implicated transmission routes and vehicles. *Epidemiol. Infect.* **2013**, *141*, 1563–1571. [CrossRef]

51. Glass, R.I.; Parashar, U.D.; Estes, M.K. Current Concepts—Norovirus Gastroenteritis. *N. Engl. J. Med.* **2009**, *361*, 1776–1785. [CrossRef]

52. Kukkula, M.; Maunula, L.; Silvennoinen, E.; Von Bonsdorff, C. Outbreak of Viral Gastroenteritis Due to Drinking Water Contaminated by Norwalk-like Viruses. *J. Infect. Dis.* **1999**, *180*, 1771–1776. [CrossRef]

53. Doyle, M.P.; Erickson, M.C. Summer meeting 2007—The problems with fresh produce: An overview. *J. Appl. Microbiol.* **2008**, *105*, 317–330. [CrossRef] [PubMed]

54. Maunula, L.; Miettinen, I.T.; Von Bonsdorff, C.H. Norovirus outbreaks from drinking water. *Emerg. Infect. Dis.* **2005**, *11*, 1716–1721. [CrossRef] [PubMed]

55. Riera-Montes, M.; Brus Sjölander, K.; Allestam, G.; Hallin, E.; Hedlund, K.O.; Löfdahl, M. Waterborne norovirus outbreak in a municipal drinking-water supply in Sweden. *Epidemiol. Infect.* **2011**, *139*, 1928–1935. [CrossRef] [PubMed]

56. Matthews, J.E.; Dickey, B.W.; Miller, R.D.; Felzer, J.R.; Dawson, B.P.; Lee, A.S.; Rocks, J.J.; Kiel, J.; Montes, J.S.; Moe, C.L.; et al. The epidemiology of published norovirus outbreaks: A review of risk factors associated with attack rate and genogroup. *Epidemiol. Infect.* **2012**, *140*, 1161–1172. [CrossRef]

57. Robilotti, E.; Deresinski, S.; Pinsky, B.A. Norovirus. *Clin. Microbiol. Rev.* **2015**, *28*, 134–164. [CrossRef]

58. Yezli, S.; Otter, J.A. Minimum Infective Dose of the Major Human Respiratory and Enteric Viruses Transmitted Through Food and the Environment. *Food Environ. Virol.* **2011**, *3*, 1–30. [CrossRef]

59. Ciocca, M. Clinical course and consequences of hepatitis A infection. *Vaccine* **2000**, *18*, S71–S74. [CrossRef]

60. Yong, H.T.; Son, R. Review Article Hepatitis A virus—A general overview. *Int. Food Res. J.* **2009**, *467*, 455–467.

61. Jacobsen, K.H.; Wiersma, S.T. Hepatitis A virus seroprevalence by age and world region, 1990 and 2005. *Vaccine* **2010**, *28*, 6653–6657. [CrossRef]

62. Yeh, H.Y.; Hwang, Y.C.; Yates, M.V.; Mulchandani, A.; Chen, W. Detection of hepatitis A virus by using a combined cell culture-molecular beacon assay. *Appl. Environ. Microbiol.* **2008**, *74*, 2239–2243. [CrossRef]

63. Borchardt, M.A.; Spencer, S.K.; Kieke, B.A.; Lambertini, E.; Loge, F.J. Viruses in nondisinfected drinking water from municipal wells and community incidence of acute gastrointestinal illness. *Environ. Health Perspect.* **2012**, *120*, 1272–1279. [CrossRef] [PubMed]

64. Hellmér, M.; Paxéus, N.; Magnius, L.; Enache, L.; Arnholm, B.; Johansson, A.; Bergström, T.; Norder, H. Detection of pathogenic viruses in sewage provided early warnings of hepatitis A virus and norovirus outbreaks. *Appl. Environ. Microbiol.* **2014**, *80*, 6771–6781. [CrossRef] [PubMed]

65. Ouardani, I.; Manso, C.F.; Aouni, M.; Romalde, J.L. Efficiency of hepatitis A virus removal in six sewage treatment plants from central Tunisia. *Appl. Microbiol. Biotechnol.* **2015**, *99*, 10759–10769. [CrossRef] [PubMed]

66. Chigor, V.N.; Sibanda, T.; Okoh, A.I. Assessment of the Risks for Human Health of Adenoviruses, Hepatitis A Virus, Rotaviruses and Enteroviruses in the Buffalo River and Three Source Water Dams in the Eastern Cape. *Food Environ. Virol.* **2014**, *6*, 87–98. [CrossRef] [PubMed]

67. Dias, J.; Pinto, R.N.; Vieira, C.B.; de Abreu Corrêa, A. Detection and quantification of human adenovirus (HAdV), JC polyomavirus (JCPyV) and hepatitis A virus (HAV) in recreational waters of Niterói, Rio de Janeiro, Brazil. *Mar. Pollut. Bull.* **2018**, *133*, 240–245. [CrossRef] [PubMed]

68. Jacobsen, K.H. *The Global Prevalence of Hepatitis A Virus Infection and Susceptibility: A Systematic Review (No. WHO/IVB/10.01)*; WHO Press: Geneva, Switzerland, 2009.

69. Gupta, E.; Ballani, N. State of the globe: Hepatitis A virus—Return of a water devil. *J. Glob. Infect. Dis.* **2014**, *6*, 57–58. [CrossRef]

70. Osuolale, O.; Okoh, A. Incidence of human adenoviruses and Hepatitis A virus in the final effluent of selected wastewater treatment plants in Eastern Cape Province, South Africa. *Virol. J.* **2015**, *12*, 1–8. [CrossRef]

71. APHA/AWWA/WEF. Standard Methods for the Examination of Water and Wastewater. 2012. Available online: https://beta-static.fishersci.com/content/dam/fishersci/en_US/documents/programs/scientific/technical-documents/white-papers/apha-water-testing-standard-methods-white-paper.pdf (accessed on 13 February 2020).

72. ISO. Microbiology of the Food Chain—Horizontal Method for Determination of Hepatitis A Virus and Norovirus Using Real-Time RT-PCR—Part 1: Method for Quantification. ISO/TS 15216-1. Available online: https://www.iso.org/standard/65681.html (accessed on 13 February 2020).

73. Fout, G.S.; Cashdollar, J.L.; Varughese, E.A.; Parshionikar, S.U.; Grimm, A.C. EPA method 1615. Measurement of enterovirus and norovirus occurrence in water by culture and RT-qPCR. I. collection of virus samples. *J. Vis. Exp.* **2015**, e52067. [CrossRef]

74. R Core Team. R: A Language and Environment for Statistical Computing. Available online: https://www.r-project.org/ (accessed on 16 July 2019).

75. Atmar, R.L.; Opekun, A.R.; Gilger, M.A.; Estes, M.K.; Crawford, S.E.; Neill, F.H.; Graham, D.Y. Norwalk virus shedding after experimental human infection. *Emerg. Infect. Dis.* **2008**, *14*, 1553–1557. [CrossRef]

76. Lee, S.G.; Jheong, W.H.; Suh, C.I.; Kim, S.H.; Lee, J.B.; Jeong, Y.S.; Ko, G.P.; Jang, K.L.; Lee, G.C.; Paik, S.Y. Nationwide groundwater surveillance of noroviruses in South Korea, 2008. *Appl. Environ. Microbiol.* **2011**, *77*, 1466–1474. [CrossRef] [PubMed]

77. Bae, J.; Schwab, K.J. Evaluation of murine norovirus, feline calicivirus, poliovirus, and MS2 as surrogates for human norovirus in a model of viral persistence in surface water and groundwater. *Appl. Environ. Microbiol.* **2008**, *74*, 477–484. [CrossRef] [PubMed]

78. Ngazoa, E.S.; Fliss, I.; Jean, J. Quantitative study of persistence of human norovirus genome in water using TaqMan real-time RT-PCR. *J. Appl. Microbiol.* **2008**, *104*, 707–715. [CrossRef] [PubMed]

79. Seitz, S.R.; Leon, J.S.; Schwab, K.J.; Lyon, G.M.; Dowd, M.; McDaniels, M.; Abdulhafid, G.; Fernandez, M.L.; Lindesmith, L.C.; Baric, R.S.; et al. Norovirus infectivity in humans and persistence in water. *Appl. Environ. Microbiol.* **2011**, *77*, 6884–6888. [CrossRef]

80. Wyer, M.D.; Wyn-Jones, A.P.; Kay, D.; Au-Yeung, H.K.C.; Gironés, R.; López-Pila, J.; de Roda Husman, A.M.; Rutjes, S.; Schneider, O. Relationships between human adenoviruses and faecal indicator organisms in European recreational waters. *Water Res.* **2012**, *46*, 4130–4141. [CrossRef]

81. Henry, R.; Schang, C.; Kolotelo, P.; Coleman, R.; Rooney, G.; Schmidt, J.; Deletic, A.; McCarthy, D.T. Effect of environmental parameters on pathogen and faecal indicator organism concentrations within an urban estuary. *Estuar. Coast. Shelf Sci.* **2016**, *174*, 18–26. [CrossRef]

82. Da Silva, A.K.; Le Saux, J.C.; Parnaudeau, S.; Pommepuy, M.; Elimelech, M.; Le Guyader, F.S. Evaluation of removal of noroviruses during wastewater treatment, using real-time reverse transcription-PCR: Different behaviors of genogroups I and II. *Appl. Environ. Microbiol.* **2007**, *73*, 7891–7897. [CrossRef]

83. Iwai, M.; Hasegawa, S.; Obara, M.; Nakamura, K.; Horimoto, E.; Takizawa, T.; Kurata, T.; Sogen, S.I.; Shiraki, K. Continuous presence of noroviruses and sapoviruses in raw sewage reflects infections among inhabitants of Toyama, Japan (2006 to 2008). *Appl. Environ. Microbiol.* **2009**, *75*, 1264–1270. [CrossRef]

84. López-Gálvez, F.; Truchado, P.; Sánchez, G.; Aznar, R.; Gil, M.I.; Allende, A. Occurrence of enteric viruses in reclaimed and surface irrigation water: Relationship with microbiological and physicochemical indicators. *J. Appl. Microbiol.* **2016**, *121*, 1180–1188. [CrossRef]

85. Haramoto, E.; Fujino, S.; Otagiri, M. Distinct behaviors of infectious F-specific RNA coliphage genogroups at a wastewater treatment plant. *Sci. Total Environ.* **2015**, *520*, 32–38. [CrossRef]

86. Haramoto, E.; Kitajima, M.; Hata, A.; Torrey, J.R.; Masago, Y.; Sano, D.; Katayama, H. A review on recent progress in the detection methods and prevalence of human enteric viruses in water. *Water Res.* **2018**, *135*, 168–186. [CrossRef] [PubMed]

 *water*

Article

Wastewater Reclamation in Major Jordanian Industries: A Viable Component of a Circular Economy

Motasem N. Saidan [1,*], **Mohammad Al-Addous** [2], **Radwan A. Al-Weshah** [3], **Ibrahim Obada** [4], **Malek Alkasrawi** [5] **and Nesrine Barbana** [6]

[1] Department of Chemical Engineering, School of Engineering, The University of Jordan, Amman 11942, Jordan
[2] Department of Energy Engineering, School of Natural Resources Engineering and Management, German Jordanian University, Amman 11180, Jordan; mohammad.addous@gju.edu.jo
[3] Civil Engineering Department, School of Engineering, The University of Jordan, Amman 11942, Jordan; weshah11@yahoo.com
[4] Ministry of Water and Irrigation, Amman 11181, Jordan; ibrahim.obadah@gmail.com
[5] Department of PS & Chemical Engineering, University of Wisconsin, Stevens Point, WI 54481, USA; malek.alkasrawi@uwsp.edu
[6] Department of Environmental Technology, Technische Universität Berlin, 10623 Berlin, Germany; nesrine@campus.tu-berlin.de
* Correspondence: m.saidan@gmail.com or m.saidan@ju.edu.jo

Received: 13 March 2020; Accepted: 27 April 2020; Published: 30 April 2020

Abstract: Water scarcity remains the major looming challenge that is facing Jordan. Wastewater reclamation is considered as an alternative source of fresh water in semi-arid areas with water shortage or increased consumption. In the present study, the current status of wastewater reclamation and reuse in Jordan was analyzed considering 30 wastewater treatment plants (WWTPs). The assessment was based on the WWWTPs' treatment processes in Jordan, the flowrates scale, and the effluents' average total dissolved solid (TDS) contents. Accordingly, 60% of the WWTPs in Jordan used activated sludge as a treatment technology; 30 WWTPs were small scale ($<1 \times 10^4$ m^3/day); and a total of 17.932 million m^3 treated wastewater had low TDS (<1000 ppm) that generally can be used in industries with relatively minimal cost of treatment. Moreover, the analysis classified the 26 million m^3 groundwater abstraction by major industries in Jordanian governorates. The results showed that the reclaimed wastewater can fully offset the industrial demand of fresh water in Amman, Zarqa, and Aqaba governorates. Hence, the environmental assessment showed positive impacts of reclaimed wastewater reuse scenario in terms of water depletion (saving of 72.55 million m^3 groundwater per year) and climate change (17.683 million kg CO_{2Eq} reduction). The energy recovery assessment in the small- and medium-scale WWTPs ($<10 \times 10^4$ m^3/day) revealed that generation of electricity by anaerobic sludge digestion equates potentially to an offset of 0.11–0.53 kWh/m^3. Finally, several barriers and prospects were put forth to help the stakeholders when considering entering into an agreement to supply and/or reuse reclaimed water.

Keywords: reclaimed water; circular economy; anaerobic digestion; biogas; reuse; water pricing; water depletion; industrial sector

1. Introduction

Water is becoming a limited resource in terms of quantity and quality due to the growing global economy and population, accelerating urbanization, and climate change effects [1–3]. Water reuse has been employed as an alternative water supply in arid and semi-arid regions [4,5]. In this context, wastewater and water reclamation plays a vital role in sustainable water resource management and mainly in various application such as agricultural irrigation, industrial processes, aquaculture, and for any non-human contact utilization, etc. [5–8]. Moreover, reclaimed wastewater is a resource that can be continuously produced unaffected by climatic conditions [9,10], especially in the Mediterranean region, one of the most vulnerable areas to climate change and with limited water resources [11–13].

However, the potential of recycling and reusing treated wastewater in a transition to a circular economy should be exploited thoroughly in arid and semi-arid areas, since it could synergize the wide adoption of water reuse as an alternate water supply [14–16].

Wastewater Reclamation Overview in Jordan

Jordan is classified as a semi-arid to arid country, with scarce water resources compared with other countries in the Middle East, and is ranked among the poorest countries in the world in terms of water availability [17–25]. Figure 1 shows the water resources in Jordan including locations of wastewater treatment plants (WWTPs). The Syria crisis is still adding strain on Jordan's economy and infrastructure and has put pressure on all sectors including water, municipal services, and electricity supply [26–34]. This problem is even more intense in areas with high population due to refugee influx that caused unsustainable over-exploitation of groundwater, and consequently led to increasing groundwater salinity and depleting resources (i.e., the water table was reduced by 5 m in areas like Dhuleil-Hallabat, area of the Amman-Zaraq basin, and tripled in salinity) [35].

Most of the published literature on water reuse in the Middle East focused on reclaimed wastewater uses in agricultural fields [36–40]. For instance, Hussain et al. (2019) reviewed 124 recent publications on the multiple aspects of safe use of treated wastewater for agriculture, landscape, and forestry and for non-conventional water resources management [40]. Moreover, it is also reported that approximately 20 million hectares of arable land worldwide is irrigated with wastewater [41].

Considering the water scarcity situation, Jordan has given top priority to the use of reclaimed wastewater in agriculture and industrial sectors [42–45], hence, the reuse of wastewater in agriculture has replaced freshwater resources, which were previously used for irrigation, allowing freshwater to be reallocated to the municipal sector where there is higher demand and quality water is needed for potable use. Despite of that, the agricultural sector accounts for 75% of all water consumption in Jordan and produces only 2% of the Gross Domestic Product (GDP) [46]. On the other hand, to reuse the reclaimed wastewater eco-efficiently in the industrial sector, most of the industrial facilities need to improve their wastewater management practices and upgrade their on-site treatment units to treat the wastewater before use [47].

The WWTPs play a vital role to decrease the environmental impacts of municipal and industrial discharges [48], while having an advanced (tertiary) treatment, the wastewater recycling and reuse can be promoted [49], as well as, enhancing the recovery of materials or energy [50]. Wastewater reclamation is one of the recommended solutions for the problem of water scarcity although the process may be complex, costly in terms of resources, and energy demanding depending on the quality of treated wastewater and the adopted technology for tertiary treatment [51]. However, shifting of the WWTP effluents from their application in agricultural irrigation to the industrial sector will require recognition of the fact that some agricultural activity would no longer have access to water for irrigation. Despite this, such a shift is recommended since most of the existing conventional treatment of WWTPs (mechanical chemical and biological treatment) does not eliminate emerging pollutants (i.e., pharmaceuticals and personal care products, hormones and steroids, persistent organic pollutants, etc.) from the wastewater, which can be induced into the food chain, subsequently causing adverse ecological and human health effects [52].

So far, wastewater reclamation and reuse in the context of water shortage in Jordan is not high, overall, whereas the potentiality of wastewater reuse is huge. The objectives of this paper are to compressively analyze the current status of wastewater reclamation and its reuse in major industries in Jordan, and to summarize the opportunities and the challenges of expanding wastewater reuse, and then to put forth prospects for future wastewater reclamation and reuse in Jordan.

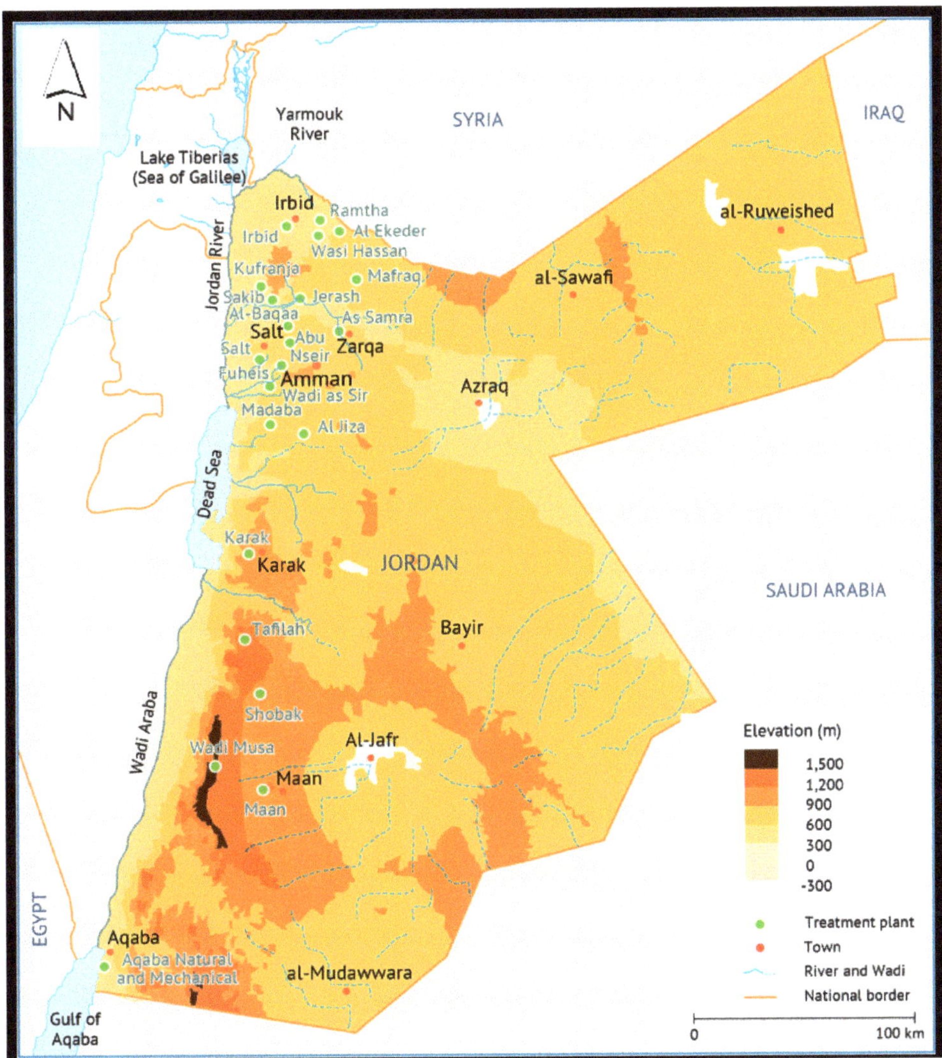

Figure 1. Water resources in Jordan.

2. Materials and Methods

2.1. Wastewater Treatment Plants in Jordan

Jordan has a fair operational capacity in wastewater treatment, although it is highly cost-intensive. The 34 central WWTPs are expected to treat 240 million m^3 per year (MCM/year) by 2025 [18]. Increasing sanitation coverage is expensive, and the proposed shift in water sector expenditures from water supply to sanitation in 2011–2013 is a significant step toward increasing coverage. In 2013, collection costs amounted to JOD 47 million (1$ is 0.71 Jordan Dinar (JOD)) and treatment costs to JOD 43.1 million [53]. Moreover, water and sanitation service costs are subsidized. Combined water and sewer bills amount to less than 0.92% of the total household annual expenditures. With Jordan's population expected to almost double by 2050, water demand will exceed the available water resources by more than 26% [18].

Figure 2 shows the variety and distribution of 34 different processes in WWTPs in Jordan. The most widely used technologies are the activated sludge (AS) process with a share of 60%. Followed by the wastewater stabilization pond (WSP) process with a share of 19%. While the trickling filter (TF) and AS process, Membrane Bioreactor (MBR) and TF process, and oxidation sludge (OS) process were evenly having the same use share of 6%, respectively. The TF process was the least used technology with a share of 3%. Moreover, one of these WWTPs is of super-large scale (>30 × 10^4 m^3/day), 4 WWTPs are of medium scale (1 × 10^4–10 × 10^4 m^3/day), and 30 WWTPs are small scale (<1 × 10^4 m^3/day), which are generally built in medium and small size cities and refugees camps.

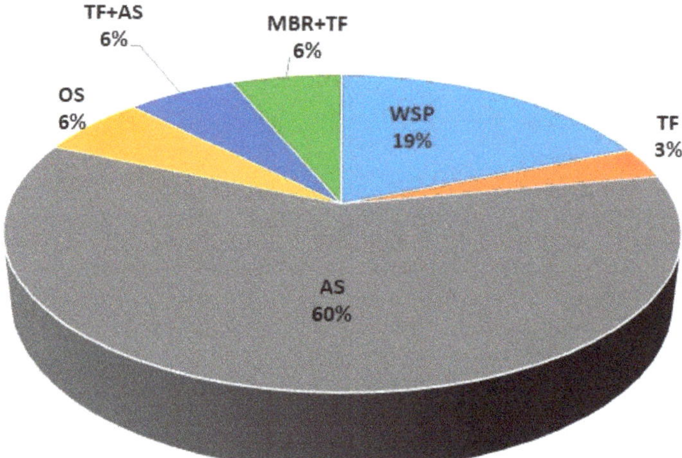

Figure 2. The variety and distribution of different processes in wastewater treatment plants (WWTPs) in Jordan. AS stands for activation sludge; OS is oxidation sludge; TF is trickling filter; WSP is wastewater stabilization pond; MBR + TF is Membrane Bioreactor and TF process; and TS + AS is trickling filter and activation sludge process.

2.2. Data Gathering and Analysis

The analysis carried out in the present study is divided into four main steps as illustrated in Figure 3, which shows the methodological approach to addressing the specific objectives of this study.

A desk study was carried out for the available baseline documents (i.e., unpublished, monthly progress reports, internal memos, and minutes of meetings) and other references for collecting the technical data. The data and information used in the present study were gathered via semi-structured interviews with key stakeholders in the water (Ministry of Water and Irrigation, Ministry of Agriculture, Ministry of Environment, etc.) and industrial sectors (Ministry of Trade and Industry, Chambers of Industry, etc.), and with international funding agencies (i.e., USAID, GIZ, etc.) involved in the ongoing projects targeting integrated water resource management in Jordan. In addition, qualitative and quantitative data and information have been derived from unpublished government reports.

Moreover, before the interviews, a brief session was hosted to probe respondents for greater clarity in answers and consistency in relation to the objectives of the questions.

Information obtained through the interviews was crosschecked with the objective to reassess gaps and divergences of information.

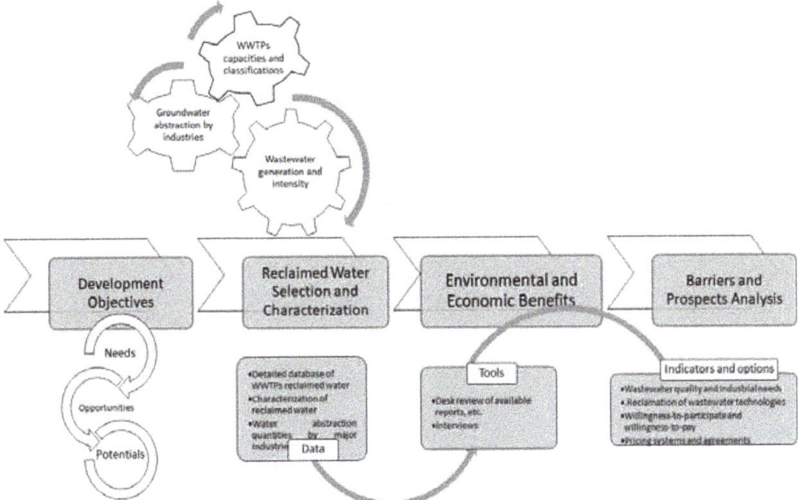

Figure 3. Diagram of the framework.

3. Results and Discussion

3.1. Wastewater Reclamation: Current Capacity and Potential Reuse

3.1.1. Reclaimed Wastewater Production: Overview and Potentials

Most of the WWTPs in Jordan provide secondary treatment with a variety of activated sludge processes followed by disinfection with chlorine. The exception is the Aqaba treatment facility, which provides tertiary filtration of the oxidized secondary effluent followed by ultraviolet disinfection and a chlorine residual. The total effluent of the wastewater flow from the WWTPs is around 166 million m^3 based on data obtained from the Ministry of Water for the year 2018, as shown in Table 1.

The industrial sector mostly relies on fresh water, which could be used for domestic purposes. For instance, the industry uses 32.2 million m^3 groundwater, 4.8 million m^3 surface water, and 1.7 million m^3 of treated wastewater [18]. Thus, this provides a great opportunity for groundwater-to-recycled water substitution.

In the present study, the WWTP effluents were classified according to their average total dissolved solids contents (TDSs) as follows: <1000 ppm; 1000 < TDS <1500; and >1500, based on wastewater analysis data (average data 2010–2016). Figure 4 shows the classification of WWTPs according to their effluents' TDS.

Table 1 shows the annual WWTP effluents' flow rate according to the TDS classifications. The first class (TDS < 1000 ppm), which relatively has the lowest TDS, can be reused several times in most industrial applications, especially in thermal units, cooling towers, etc. For instance, Aqaba recycled water, which has the lowest salinity among the WWTPs in Jordan (TDS = 587 ppm), is most readily usable in industrial applications. So potentially, this class represents 9 WWTPs distributed in different locations in Jordan, as shown in Table 1, and, in total, 17.932 million m^3 of treated wastewater of this class can be used directly with no or low cost of on-site treatment in the industrial sector depending on the fit-for-purpose water criteria.

However, the second class (1000 < TDS < 1500), which has medium TDS, has the highest annual effluent flow rate of 147.323 million m^3 in total out of 18 WWTPs distributed in widely different locations in Jordan, as shown in Table 1. The most effluent wastewater flowrate in this class is generated from Al Samra WWTP with 117.1 million m^3 per year by offering sanitation services to about two million in Amman and Zarqa governorates; the first and third most populated cities in

Jordan, respectively [44,54]. With such large capacity and modern technology to ensure the highest purifications, Al Samra is considered as one of the largest plants in the region [40], which treats about 70.54% of total reclaimed wastewater in Jordan. This class represents 18 WWTPs distributed in different locations in Jordan, and, in total, 147.33 million m^3 of treated wastewater of this class can be used with medium cost of some necessary modification in the plant process in the industrial sector depending on the fit-for-purpose water criteria.

The third class has a TDS > 1500, the WWTP effluents in this class cannot be used without further intensive treatment such as: demineralization; blending with low-salinity water; and some change in the industrial process. This class represents three WWTPs with 0.788 million m^3 of treated wastewater, as shown in Table 1. Therefore, due to the high capital cost of investment and relatively expensive operating cost, this class is excluded from the present study analysis.

Table 1. Annual WWTP effluent flow rate according to the total dissolved solids (TDSs) classifications.

TDS Classification	WWTP	Effluent Flow Rate (Million m^3/year)	Total Flowrate of Grouped WWTPs (Million m^3/year)
<1000 ppm	Aqaba-Tertiary	3.90	17.93
	Aqaba-Lagoon	2.22	
	Wadi Essir	1.71	
	Wadi Musa	1.02	
	Salt	3.19	
	Fuhis	1.15	
	Abu Nusseir	1.31	
	Madaba	2.53	
	Ma'an	0.92	
1000 < TDS < 1500	Karak	0.54	147.32
	Mafraq	1.29	
	Mu'taa	0.55	
	Wadi Hassan	0.38	
	Al Samra	117.10	
	Irbid	3.10	
	Wadi Shalallah	3.43	
	Kufranja	1.25	
	Jeza	0.29	
	South Amman	4.72	
	Tafileh	0.80	
	Wadi Arab	4.98	
	Ain Albasha	5.12	
	Al-Me'rad	1.16	
	North Shouneh	0.15	
	Tal-Almanttah	0.15	
	Akeeder	0.82	
	Ramtha	1.50	
>1500	Jerash	0.42	0.78
	Shobak	0.05	
	Al Lujjon	0.30	

Excluding food and pharmaceutical industries, the total groundwater abstraction for industrial purposes was approximately 26 million m^3 in Jordan in 2015 [55]. The major industries considered in the present study as the major groundwater abstracting industries are clarified in Table 2. Considering this, Figure 5 shows the total groundwater abstraction by major industries in Jordanian governorates, where the industries in Karak governorate were the most groundwater abstracting, with approximately 11.5 million m^3 per year. Followed by the industries in Ma'an (4.38 million m^3 per year). While the industries in Zarqa and Aqaba governorates were close to each other in terms of groundwater abstraction with 2.74 and 1.825 million m^3 per year, respectively. The industries in the Capital Amman were the least groundwater abstracting with 0.611 million m^3 per year.

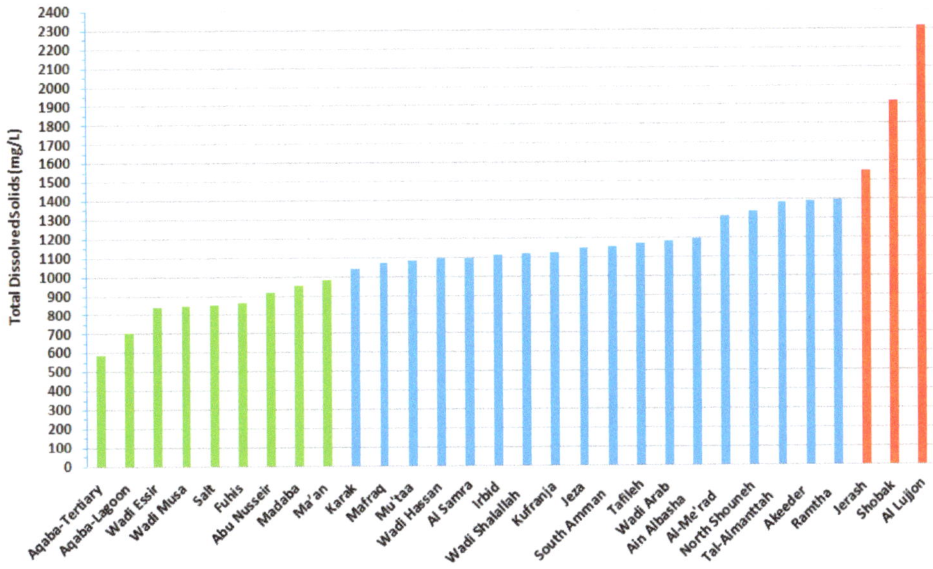

Figure 4. Classification of WWTPs according to their effluents' TDS.

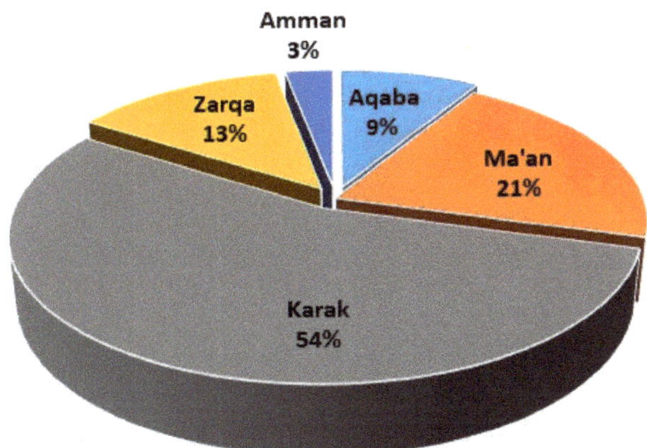

Figure 5. Percentages of total groundwater abstraction by major industries in Jordanian governorates.

Table 2. Major groundwater abstraction industries in Jordan according to governorates.

Governorate	Major Groundwater Abstraction Industries
Amman	Cement, metals, concrete, paper and carton, etc.
Zarqa	Petroleum refinery, metals and pipes, paper and cardboard, thermal power and electricity plants, cement, etc.
Karak	Phosphate mines, potash, chemical fertilizers, cement, mining, etc.
Ma'an	Phosphate, cement, etc.
Aqaba	Phosphate, fertilizers, etc.

Hence, based on the data of first class and second class in Table 1, the potential reclaimed wastewater substitution in major industries in Jordanian governorates is shown in Figure 6. It is obvious that the reclaimed wastewater in Zarqa governorate can fully substitute the industrial demand of fresh water (Figure 6a) and the needs for irrigation of 3000 donums for 20–30 farmers adjacent to Al Samra WWTP as reported by Hussein (2018) [54] and Maldonado (2017) [44]. The full substitution of industrial demand is also noticed in both Amman and Aqaba governorates with 13.13- and 3.36-fold, respectively. However, the shortage of industrial demand substitution is significantly clear in both of Ma'an and Karak governorates with substitution amounts of 2.45 and 10.4 million m^3 per year, respectively, as clearly shown in Figure 6b. Therefore, for the WWTPs in the governorates with a substitution factor less than one (mainly Ma'an and Karak governorates) it is preferable to prioritize their effluents (reclaimed wastewater) for irrigation use where applicable.

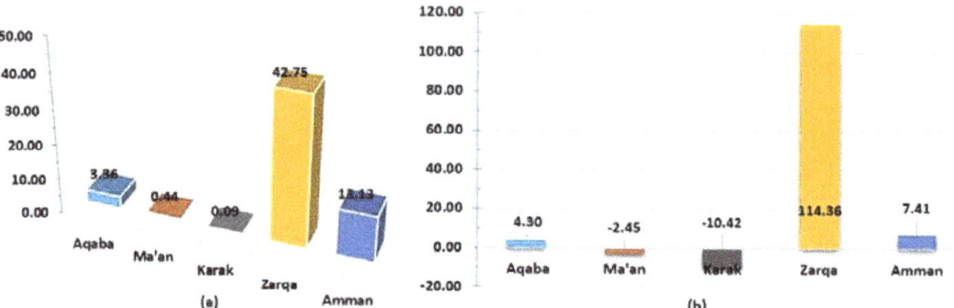

Figure 6. Reclaimed wastewater substitution in major industries in Jordanian governorates: (**a**) substitution factors, and (**b**) substitution amounts (million m^3 per year).

Figure 7 shows the responses of the interviewed industries (17 samples from those shown in Table 2). It is drastically indicated that low TDS (water salinity) is the major requirement that was requested by 35% of the responses. Interestingly, the sample responses showed willingness to accept to replace the groundwater with reclaimed wastewater.

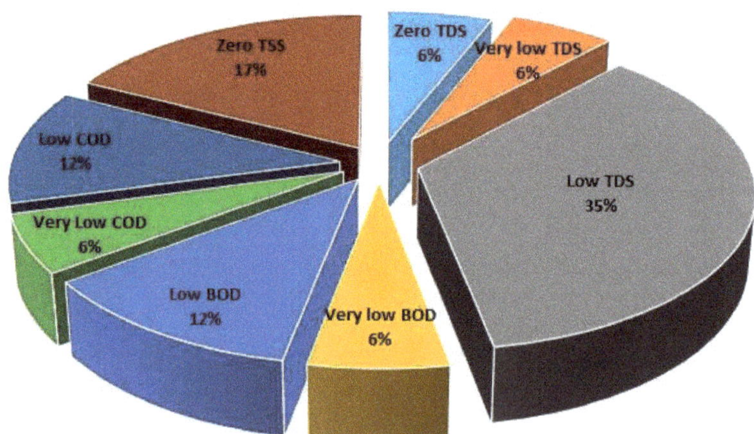

Figure 7. Responses of reclaimed wastewater quality requirements by the major industries.

However, 6% of the responses requested advanced treatment to receive very low values of TDS, biological oxygen demand (BOD), and chemical oxygen demand (COD). Zero total suspended solids (TSS) was requested by 17% of the responses, and this was mainly required for the cooling of power generators.

3.1.2. Environmental and Economic Benefits

Pintilie et al. (2016) studied the life cycle assessment (LCA) of substituting fresh water with treated wastewater obtained from tertiary treatment and concluded that it does not lead to a substantial improvement of environmental impact for most of the indicators [48]. However, only water depletion (WD) and climate change (CC) were considered in the present study to compare the environmental impact between reclaimed wastewater reuse and no reuse scenarios. WD is recommended for water-stressed situations because a net saving of water from nature represents the most important effect of water reuse. The WD indicator values proposed by Pintilie et al. (2016) were considered in the present assessment as the following: 5.74×10^{-4} m^3 per m^3 entering the whole system for the no reuse scenario, and -4.39×10^{-1} m^3 per m^3 entering the whole system for the reclaimed wastewater reuse scenario [48]. Negative values mean benefits to the environment, and positive values mean damages. Accordingly, using the data in Table 1, the annual wastewater effluent amounts (mainly the total flowrates of grouped WWTPs (million m^3/year)) of both TDS less than 1000 ppm and 1000 < TDS < 1500 ppm were 17.93 and 147.33 million m^3 per year, respectively. The sum of them is 165.26 million m^3 per year, and using the aforementioned WD indicators, the analysis revealed that 94,860 m^3 of fresh water are depleted for the scenario of no-reuse of reclaimed wastewater; however, 72.55 million m^3 of water can be saved in reclaimed wastewater reuse in major industries in Jordan, as shown in Figure 8. Results of a similar tendency were founded in literature [48,56].

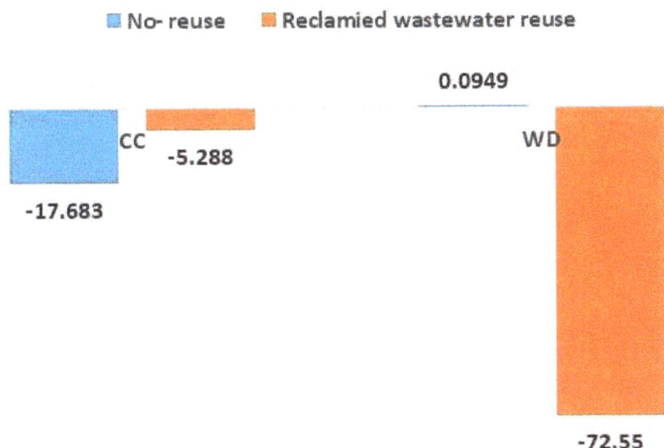

Figure 8. Water depletion (million m^3 of water) and climate change (million kg CO$_{2Eq}$) analysis for both reclaimed wastewater reuse and no reuse scenarios (negative values mean benefits to the environment, and positive values mean damages).

The CC indicator values proposed by Pintilie et al. (2016) were considered as stated above [48]. The CC indicators were with negative values (indicates benefits to the environment) according to Pintilie et al. (2016) are the following: -1.07×10^{-1} and -3.20×10^{-2} kg CO$_{2Eq}$ per m^3 reclaimed wastewater for both scenarios of reuse and no reuse, respectively [48]. Accordingly, using the data in Table 1, the annual wastewater effluent amounts (mainly the total flowrates of grouped WWTPs (million m^3/year)) of both TDS less than 1000 ppm and 1000 < TDS < 1500 ppm were 17.93 and 147.33 million m^3 per year, respectively. The sum of them is 165.26 million m^3 per year, and using the aforementioned CC indicators, as shown in Figure 8, both scenarios showed beneficial impacts (negative values) to the environment in terms of climate change impacts. The no reuse scenario has relatively higher benefits with 17.683 million kg CO$_{2Eq}$ reduction compared to a 5.288 million kg CO$_{2Eq}$ reduction for the reuse scenario.

Normally, several factors influence the reclaimed wastewater provision and exploitation as a substitute [57,58]. According to the economic analysis of wastewater reclamation in Jordan, the difference between water price and reclaimed wastewater price plays a vital role in the willingness of the industries to accept the reclaimed wastewater as substitute. Therefore, for the low TDS (<1000) reclaimed wastewater (Table 1), the average cost of one m^3 of reclaimed wastewater is estimated at

0.55 JOD (including the pipeline installation, pumping electricity, and operation and naintenance (O&M) costs), while the cost of fresh water is 1 JOD/m^3. In this case, the reclaimed wastewater is competitive to some extent with regard to its price advantage. Moreover, based on experts' estimation, the environmental value of groundwater saved in the groundwater aquifer is 1.5 JOD/m^3. Hence, the cost–benefit analysis of this case (water of TDS < 1000) is attractive for the consumer and the government.

While for reclaimed wastewater with TDS higher than 1000 ppm, a treatment is needed based on the application. Therefore, excluding the reuse of reclaimed in cement and concrete industries, the average cost of one m^3 of reclaimed wastewater is estimated at 2 JOD (including treatment, pipeline installation, pumping electricity, and O&M costs). It is worth mentioning that the long-distance pipelines from WWTPs to industrial zones and clusters, were the major cause for such costly per m^3 water cost, especially in southern Jordan clusters. In order to overcome the hesitance of industries to reuse reclaimed wastewater when advanced treatment is required, subsidies by way of discounted cost of water should be provided in addition to fund allocation for capital cost coverage when on-site treatment is needed, as well as policy reforms to enhance the financial sustainability of the water sector.

3.2. Energy Recovery from Wastewater Reclamation

Wastewater treatment in WWTPs (mainly AS treatment process) requires around 0.38–2.74 kWh/m^3 in Jordan, as shown in Figure 9. Additionally, 0.95–1.25 kWh/m^3 is needed for wastewater as reported in literature [59,60]. The difference in energy use needed for wastewater reclamation and supply can be reduced by recovering organic energy during the wastewater treatment process [59]. Currently, only in the Al Samra WWTP, biogas production from sludge treatment is undertaken in Jordan. As shown in Figure 10, the two types of thickened sludge are mixed in two covered tanks of 98 m^3 volume before being pumped and introduced in seven anaerobic digesters of a capacity of 15,900 m^3 each. In the digesters, the sludge is mixed thoroughly by Cannon®mixers (Trevose, PA USA) using the recycled compressed biogas. The sludge stays for three weeks at 35 °C in the digesters. Heating is done by hot water recovered from the cooling of the engines in a shell-and-tube heat exchanger. Through hydro energy and biogas production, the Al Samra WWTP has a potential energy recovery of 95% of its needs, only 5% is drawn from the national grid. Moreover, 300,000 tons of CO_2 is saved per year through energy recovery and renewable energy utilization [61].

The introduction of anaerobic sludge digestion is generally expected to offset 25–50% of an aerobic wastewater treatment plant's energy needs [59,63,64], however, based on WWTP data gathered in Jordan, having anaerobic sludge digestion in the small- and medium-scale WWTPs (<10 × 10^4 m^3/day) can potentially produce electricity that would equate to an offset of 0.11–0.53 kWh/m^3. Consequently, this may help in reducing the costs of reclaimed wastewater reuse with further treatment requirements mainly for reclaimed wastewater with TDS higher than 1000 ppm as stated before.

However, energy produced from anaerobic sludge digestion can be feasibly increased by co-digestion with kitchen or other organic wastes [65–69]. Currently, the co-digestion is only applied at a laboratory scale in Jordan. Al-Addous et. al. (2019) evaluated the potential biogas production from the co-digestion of municipal food waste and wastewater sludge at a refugee camp. Accordingly, a possible ratio to start with is 60–80% organic waste, which can produce 21–65 m^3 biogas ton^{-1} of fresh matter [70].

Notwithstanding that co-digestion does not exist in Jordan yet, the anaerobic digestion systems tend to be well operated in Jordan (i.e., Al Samra WWTP). Hence, when co-digestion is utilized in Jordan, this will be a vital opportunity to make cost-effective use of existing facilities and improve sludge biogas potential [71–74].

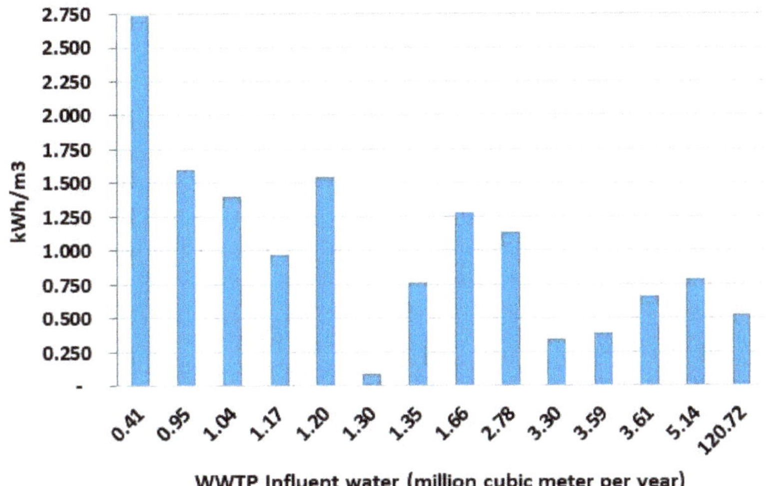

Figure 9. WWTPs' electricity consumption data for the activated sludge (AS) treatment process.

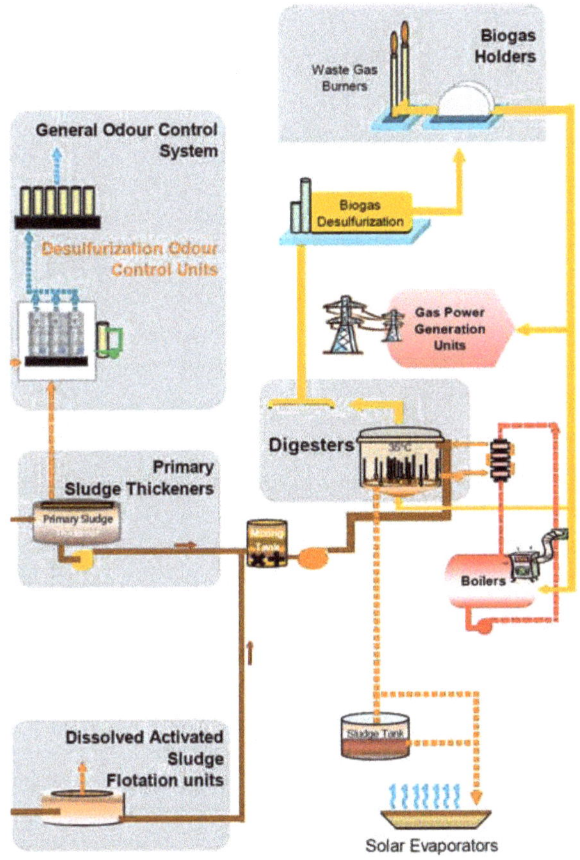

Figure 10. Biogas production in anaerobic digesters at the Al Samra WWTP [62].

3.3. Reclaimed Wastewater Reuse: Barriers and Prospects

3.3.1. Reclaimed Wastewater Quality and Industrial Needs

Industrial uses of reclaimed wastewater come in many different ways such as cooling-water, processing, and boiler feed water. Therefore, the process water requirements for water quality vary depending on the industry. Some of the concerns for industrial use of reclaimed wastewater are corrosion, scaling, and biological growth; however, these concerns are applicable to potable water as well. Most of cooling water is treated already to address these concerns. For instance, corrosion is a concern in cooling water no matter whether the facility uses potable water or reclaimed wastewater. Scaling from dissolved minerals such as calcium, magnesium, and phosphates can be controlled by monitoring and chemically treating the water to prevent scaling. Magnesium–phosphorus precipitation from sludge and the recovery of struvite after anaerobic sludge treatment process is conducted in order to prevent clogging in pumps and pipes in any further reuse applications [75–78]. Biological concerns can be addressed by adding chlorine to levels of 2.0 mg/L that will kill most microorganisms that causes corrosion or deposits in cooling systems [79].

To facilitate the use of recycled water in industrial applications, the information on the quality of the municipal recycled water should be provided and available to the industrial users. Moreover, opportunities to improve water quality for specific purposes, either by the supplier through additional treatment and/or source control, or the industrial user can improve treatment and control processes to levels specific to its process needs.

3.3.2. Reclamation of Wastewater Technologies

Lyu et al. (2016) discussed all advances in technology by which wastewater may be treated to meet the most stringent quality requirements and be used for any purposes desired [49]. For instance, the technologies applied in wastewater reuse include: (1) oxidants for disinfection purposes using sodium hypochlorite; ultraviolet radiation [80]; and ozone for high bactericidal disinfection [81] and the removal of 90–99% for antibiotics and estrogens [82]; (2) biological treatments such as anaerobic, maturation ponds and constructed wetlands [83–85]; (3) physical separations such as membrane filtration for 81% removal of electroconductivity, 83% for Na$^+$, and 80% for Cl$^-$ [86]; the removal of 95% of heavy metals [87]; the removal of >89% of pharmaceuticals [88]; (4) electrochemical treatments to completely remove *Escherichia coli* [89–91]; as well as, (5) solar photocatalysis with TiO$_2$ for >90% removal of emerging pollutants (i.e., pharmaceuticals and personal care products) [92], and removals of 33% for Cd and 75% for Co [93].

3.3.3. Reclaimed Wastewater Supply Continuity

The industry demands a constant non-interrupted flow of reclaimed wastewater throughout the day [94]. In Jordan, although the reclaimed wastewater supply volumes vary diurnally and seasonally, its continuity is not so critical since WWTP effluents have relatively uninterruptable higher flows than the demand flows needed by the nearby main industries. It is noteworthy that flow equalization and water conveying capacities should be investigated to match the supplies with the demands and vice versa [95].

3.3.4. Willingness to Participate and Willingness to Pay

As deduced based on the results of these interviews, most of main industries considered in the present study expressed a positive stance toward reclaimed wastewater reuse, while they are willing to pay a significantly less amount of money than they already pay, for freshwater. Therefore, a comprehensive survey about the willingness of the industrial sector to switch to the use of reclaimed wastewater instead of groundwater is of high significance. Such surveys will help in providing more accurate data for the financial evaluation of the recycled water service and a basis for negotiation with the industries.

Factors that influence industrial user's 'willingness to pay' for reclaimed wastewater include: (1) price of alternative water sources (i.e., potable, surface water, and groundwater supplies); (2) perception of the scarcity of alternative sources; (3) capital and operating costs of switching to reclaimed wastewater supply; (4) reclaimed wastewater quality, quantity, and levels of service and reliability of supply.

3.3.5. Pricing Systems

A range of pricing systems for reclaimed wastewater can be proposed in Jordan and assessed on a win–win situation. The pricing systems can be employed alone or in combination [96], which are, but not limited to: (1) A usage fee scheme in which the industries finance the infrastructure installation, and then the usage charge offsets the supply cost of the reclaimed wastewater. For instance, such type of pricing was adopted in 2003 by the Australian government under the national water reform process [97]. (2) A connection fee which is a once-off contribution toward the cost of infrastructure needed to deliver reclaimed wastewater to the industry's delivery point. This fee may be subject to negotiation between the supplier and the industries to agree on a financial arrangement where both parties may fully or partially cover the fee of the actual work to deliver the reclaimed water to the delivery point. (3) A flat fee regardless of use ("take or pay" arrangement). For instance, regardless of actual use, the industries are obliged to pay for 75–100% of the contracted recycled water volume, and for all water consumed by the industries above the contracted level. Although this pricing scheme provides the WWTPs with guaranteed income that sustains the financials of running the scheme, it may encourage overuse of reclaimed wastewater by the industry and improper discharges to the environment.

3.3.6. Reclaimed Wastewater Agreements

Specific reclaimed wastewater guidelines are important in managing the supply and use of reclaimed wastewater particularly in relation to quantity and quality [98]. Through the agreement negotiations between the supplier of reclaimed wastewater and the customers (i.e., industries). Wherein, the parties agree to a set of obligations and responsibilities under which the reclaimed wastewater reuse scheme will operate [99]. Key issues that reclaimed wastewater agreements should cover include: (a) price, quantity, and quality of reclaimed wastewater; (b) security of the reclaimed wastewater supply; (c) measures to identify, allocate, and manage risks and ensure safe use of reclaimed wastewater; (d) liabilities and insurance for potential damages caused by supply and use; and (e) compliance with legislative and common law requirements.

4. Conclusions

The following findings can be concluded in the present study:

- Jordan is classified as a semi-arid to arid country and is ranked among the poorest countries in the world in terms of water availability. Therefore, reclaimed wastewater reuse has been driven as an alternative water supply in such looming challenges of water scarcity. For instance, a total of 26 million m^3 of groundwater abstraction is exploited annually for industrial purposes.
- In the present study, the 34 processes in WWTPs in Jordan were assessed in terms of their treatment processes, scale, and effluent TDS. The most widely used technologies are AS (60%) and WSP (19%), while the TF and AS process, MBR and TF process, and OS processes were had an even use share of 6% each. Moreover, 30 WWTPs were classified as small scale ($<1 \times 10^4$ m^3/day), which were generally built in medium- and small-size cities and refugee camps. Moreover, the analysis showed that 17.932 million m^3 of treated wastewater has low TDS < 1000 ppm and can be reused several times in most industrial applications, especially in thermal units, cooling towers, etc. However, highest annual effluents flow rate of 147.323 million m^3 in total out of 18 WWTPs distributed in widely different locations in Jordan have 1000 < TDS < 1500, which can be used with medium cost depending on the fit-for-purpose water criteria.

- Full substitution of industrial demand by reclaimed wastewater reuse can be achieved in both Amman and Aqaba governorates with 13.13- and 3.36-fold, respectively. However, the shortage of industrial demand substitution by reclaimed wastewater is significantly clear in both of Ma'an and Karak governorates with substitution amounts of 2.45 and 10.4 million m^3 per year, respectively.

- The environmental assessment showed positive impacts of reclaimed wastewater reuse scenario in terms of water depletion (saving of 72.55 million m^3 of groundwater per year) and climate change (17.683 million kg CO_{2Eq} reduction).

- From circular economic perspective, and based on WWTP data gathered in Jordan, having anaerobic sludge digestion in the small- and medium-scale WWTPs ($<10 \times 10^4$ m^3/day) can potentially produce electricity that would equate to an offset of 0.11–0.53 kWh/m^3. Consequently, this may help in reducing the costs of reclaimed wastewater reuse with further treatment requirements mainly for reclaimed wastewater with TDS higher than 1000 ppm as stated before.

- It is recommended in the present study that reclaimed wastewater agreement negotiations should be promoted between the supplier of reclaimed wastewater and the customers (i.e., industries). Moreover, indicators such as willingness to participate and willingness to pay need to be significantly determined in order to reach a win–win scheme of reclaimed water pricing model.

Author Contributions: Conceptualization, M.N.S. and I.O.; Methodology, M.N.S., I.O., and R.A.A.-W.; Formal analysis, M.N.S., M.A., M.A.-A., N.B., and I.O.; Investigation, M.N.S., R.A.A.-W., and I.O.; Resources, M.N.S. and I.O.; Writing—original draft preparation, M.N.S., M.A., M.A.-A., and I.O.; Writing—review and editing, M.N.S. and R.A.A.-W.; Project administration, M.N.S., N.B. and I.O. All authors have read and agreed to the published version of the manuscript.

Funding: This research received no external funding.

Acknowledgments: The corresponding author does acknowledge with gratitude the University of Jordan for granting him sabbatical assistance to undertake this research study. The authors also thank the anonymous reviewers and the editor of the journal for their constructive feedback.

Conflicts of Interest: The authors declare no conflict of interest.

References

1. Forslund, A.; Malm Renöfält, B.; Barchiesi, S.; Cross, K.; Davidson, S.; Ferrel, T.; Korsgaard, L.; Krchnak, K.; McClain, M.; Meijer, K.; et al. *Securing Water for Ecosystems and Human Well-being: The Importance of Environmental Flows (Swedish Water House Report 24)*; Swedish International Water Institute: Stockholm, Sweden, 2009.

2. Roccaro, P.; Verlicchi, P. Wastewater and reuse. *Curr. Opin. Environ. Sci. Health* **2018**, *2*, 61–63. [CrossRef]

3. Sun, Y.; Chen, Z.; Wu, G.; Wu, Q.; Zhang, F.; Niu, Z.; Hu, H.Y. Characteristics of water quality of municipal wastewater treatment plants in China: Implications for resources utilization and management. *J. Clean. Prod.* **2016**, *131*, 1–9. [CrossRef]

4. Shannon, M.; Bohn, P.W.; Elimelech, M.; Georgiadis, J.G.; Mariñas, B.J.; Mayes, A.M. Science and technology for water purification in the coming decades. *Nature* **2008**, *452*, 301–310. [CrossRef] [PubMed]

5. Akhoundi, A.; Nazif, S. Sustainability assessment of wastewater reuse alternatives using the evidential reasoning approach. *J. Clean. Prod.* **2018**, *195*, 1350–1376. [CrossRef]

6. Chen, W.; Geng, Y.; Dong, H.; Tian, X.; Zhong, S.; Wu, Q.; Li, S. An emergy accounting based regional sustainability evaluation: A case of Qinghai in China. *Ecol. Indic.* **2018**, *88*, 152–160. [CrossRef]

7. Yan, H.; Wu, F.; Dong, L. Influence of a large urban park on the local urban thermal environment. *Sci. Total Environ.* **2018**, *622*, 882–891. [CrossRef]

8. Zhou, Z.; Zhang, X.; Dong, W. Fuzzy comprehensive evaluation for safety guarantee system of reclaimed water quality. *Procedia Environ. Sci.* **2013**, *18*, 227–235. [CrossRef]

9. Ait-Mouheb, N.; Bahri, A.; Thayer, B.B.; Benyahia, B.; Bourrié, G.; Cherki, B.; Condom, N.; Declercq, R.; Gunes, A.; Héran, M.; et al. The reuse of reclaimed water for irrigation around the Mediterranean Rim: A step towards a more virtuous cycle? *Reg. Environ. Change* **2018**, *8*, 693–705. [CrossRef]

10. Licciardello, F.; Milani, M.; Consoli, S.; Pappalardo, N.; Barbagallo, S.; Cirelli, G. Wastewater tertiary treatment options to match reuse standards in agriculture. *Agric. Water Manag.* **2018**, *210*, 232–242. [CrossRef]

11. Collet, L.; Denis, R.; Borrell Estupina, V.; Dezetter, A.; Servat, E. Water supply sustainability and adaptation strategies under anthropogenic and climatic changes of a meso-scale Mediterranean catchment. *Sci. Total Environ.* **2015**, *536*, 589–602. [CrossRef]

12. La Jeunesse, I.; Cirelli, C.; Aubin, D.; Larrue, C.; Sellami, H.; Afifi, S.; Bellin, A.; Benabdallah, S.; Bird, D.N.; Deidda, R.; et al. Is climate change a threat for water uses in the Mediterranean region? Results from a survey at local scale. *Sci. Total Environ.* **2016**, *543*, 981–996. [CrossRef] [PubMed]

13. European Environmental Agency. *Climate Change, Impacts and Vulnerably in Europe—An Indicator-Based Report*; European Environmental Agency (2012) EEA report no 12/2012; European Environmental Agency: København, Denmark, 2012.

14. Voulvoulis, K. Water reuse from a circular economy perspective and potential risks from an unregulated approach. *Curr. Opin. Environ. Sci. Health* **2018**, *2*, 32–45. [CrossRef]

15. Coats, E.R.; Wilson, P.I. Toward nucleating the concept of the water resource recovery facility (WRRF): Perspective from the principal actors. *Environ. Sci. Technol.* **2017**, *51*, 4158–4164. [CrossRef]

16. Sgroi, M.; Vagliasindi, F.G.A.; Roccaro, P. Feasibility, sustainability and circular economy concepts in water reuse. *Curr. Opin. Environ. Sci. Health* **2018**, *2*, 20–25. [CrossRef]

17. Matouq, M.; El Hasan, T.; Al Bilbisi, H.; Abdelhadi, M.; Hindiyeh, M.; Eslaiman, S.; Duheisat, S. The climate change implication on Jordan: A case study using GIS and artificial neural networks for weather casting. *J. Sci. Taibah Univ.* **2013**, *7*, 44–55. [CrossRef]

18. Ministry of Water and Irrigation. *Jordan's Water Strategy 2016–2025*; Ministry of Water and Irrigation: Amman, Jordan, 2016.

19. Aboelnga, H.; Saidan, M.; Al-Weshah, R.; Sturm, M.; Ribbe, L.; Frechen, F.B. Component analysis for optimal leakage management in Madaba, Jordan. *J. Water Supply Res. Technol.* **2018**, *67*, 384–396. [CrossRef]

20. Saidan, M.N.; Al-Weshah, R.A.; Obada, I. Potential Rainwater Harvesting: Adaptation Measure for Urban Areas in Jordan. *Am. Water Works Assoc.* **2015**, *107*, 594–602. [CrossRef]

21. Al-Weshah, R.; Saidan, M.; Al-Omari, A. Environmental ethics as a tool for sustainable water resource management. *Am. Water Works Assoc.* **2016**, *108*, 175–181. [CrossRef]

22. Al-Awad, T.K.; Saidan, M.N.; Gareau, B.J. Halon management and ozone-depleting substances control in Jordan. *Int. Environ. Agreem.* **2018**, *18*, 391–408. [CrossRef]

23. Aldayyat, E.; Saidan, M.N.; Abu Saleh, M.A.; Hamdan, S.; Linton, C. Solid Waste Management in Jordan: Impacts and Analysis. *J. Chem. Technol. Metall.* **2019**, *54*, 454–462.

24. Alrabie, K.; Saidan, M.N. A preliminary solar-hydrogen system for Jordan: Impacts assessment and scenarios analysis. *Int. J. Hydrog. Energy* **2018**, *43*, 9211–9223. [CrossRef]

25. Khasawneh, H.; Saidan, M.; Al-Addous, M. Utilization of hydrogen as clean energy resource in chlor-alkali process. *Energy Explor. Exploit.* **2019**, *37*, 1053–1072. [CrossRef]

26. Saidan, M. Cross-sectional survey of non-hazardous waste composition and quantities in industrial sector and potential recycling in Jordan. *Environ. Nanotech. Monitor. Manag.* **2019**, *12*, 100227. [CrossRef]

27. Al-Addous, M.; Saidan, M.; Bdour, M.; Dalalah, Z.; Albatayneh, A.; Class, C.B. Key aspects and feasibility assessment of a proposed wind farm in Jordan. *Int. J. Low-Carbon Technol.* **2019**. [CrossRef]

28. Saidan, M.N.; Abu Drais, A.; Linton, C.; Hamdan, S. Solid waste Characterization and recycling in Syrian refugees hosting communities in Jordan. In *Waste Management in MENA Regions*, 1st ed.; Negm, A.M., Noama, E., Eds.; Earth and Environmental Sciences Series, Applied Environmental Science and Engineering for a Sustainable Future; Springer International Publishing AG, Part of Springer Nature: Cham, Switzerland, 2019.

29. Saidan, M.N.; Ansour, L.M.; Saidan, H. Management of Plastic Bags Waste: An Assessment of Scenarios in Jordan. *J. Chem. Technol. Metall.* **2017**, *52*, 148–154.

30. Hararah, M.A.; Saidan, M.N.; Alhamamre, Z.; Abu-Jrai, A.M.; Alsawair, J.; Damra, R.A. The PCDD/PCDF emission inventory in Jordan: Aqaba city. *J. Chem. Technol. Metall.* **2016**, *51*, 112–120.

31. Saidan, M.; Tarawneh, A. Estimation of potential E-waste generation in Jordan. *Ekoloji* **2015**, *24*, 60–64. [CrossRef]

32. Tarawneh, A.; Saidan, M. Households awareness, behaviors, and willingness to participate in E-waste management in Jordan. *Int. J. Ecosys* **2013**, *3*, 124–131.

33. Saidan, M.; Khasawneh, H.J.; Tayyem, M.; Hawari, M. Getting energy from poultry waste in Jordan: Cleaner production approach. *J. Chem. Technol. Metall.* **2017**, *52*, 595–601.

34. Hindiyeh, M.; Altalafha, T.; Al-Naerat, M.; Saidan, H.; Al-Salaymeh, A.; Sbeinati, L.A.; Saidan, M.N. Process modification of pharmaceutical tablet manufacturing operations: An Ecoefficiency approach. *Processes* **2018**, *6*, 15. [CrossRef]

35. METAP-World Bank. Water Quality Management Jordan—World Bank. Available online: http://siteresources. worldbank.org/EXTMETAP/Resources/WQM.JordanP.pdf (accessed on 2 October 2019).

36. Salahat, M.A.; Al-Qinna, M.I.; Badran, R.A. Potential of Treated Wastewater Usage for Adaptation to Climate Change: Jordan as a Success Story. In *Water Resources in Arid Areas: The Way Forward. Springer Water*; Abdalla, O., Kacimov, A., Chen, M., Al-Maktoumi, A., Al-Hosni, T., Clark, I., Eds.; Springer: Cham, Switzerland, 2017.

37. Carr, G.; Potter, R.B.; Nortcliff, S. Water reuse for irrigation in Jordan: Perceptions of water quality among farmers. *Agric. Water Manag.* **2011**, *98*, 847–854. [CrossRef]

38. Scott, C.A.; El-Naser, H.; Hagan, R.E.; Hijazi, A. Facing Water Scarcity in Jordan Reuse, Demand Reduction, Energy, and Transboundary Approaches to Assure Future Water Supplies. *Water Int.* **2003**, *28*, 209–216. [CrossRef]

39. The World Bank. *Water Reuse in The Arab World, from Principle to Practice, A Summary of Proceedings, Expert Consultation Wastewater Management in the Arab World 22–24 May 2011*; The World Bank: Dubai, UAE, 2011.

40. Hussain, M.I.; Muscolo, A.; Farooq, M.; Ahmad, W. Sustainable use and management of non-conventional water resources for rehabilitation of marginal lands in arid and semiarid environments. *Agric. Water Manag.* **2019**, *221*, 462–476. [CrossRef]

41. Mateo-Sagasta, J.; Medlicott, K.; Qadir, M.; Raschid-Sally, L.; Drechsel, P.; Liebe, J. *UN-Water Project on the Safe Use of Wastewater in Agriculture*; Jens Liebe, J., Ardakanian, K., Eds.; UN: Now York, NY, USA, 2013.

42. Hussein, I.; Abu-Sharar, T.M.; Al-Jayyousi, O. The use of treated wastewater for irrigation in Jordan: Opportunities and constraints. *Chart. Inst. Water Environ. J.* **2003**, *17*, 232–238.

43. Qadir, M.; Bahri, A.; Sato, T.; Al-Karadsheh, E. Wastewater production, treatment, and irrigation in Middle East and North Africa. *Irrig. Drain. Syst.* **2010**, *24*, 37–51. [CrossRef]

44. Maldonado, P.M.A. Water Resources Assessment under Semi-arid Conditions—Modelling Applications in a Complex Surface-groundwater System in Jordan. Ph.D. Thesis, KIT-Bibliothek, Karlsruhe, Germany, 2017.

45. Mohsen, M.; Jaber, J.O. Potential of industrial wastewater reuse. *Desalination* **2002**, *152*, 281–289. [CrossRef]

46. Schyns, J.F.; Hamaideh, A.; Hoekstra, A.Y.; Mekonnen, M.M.; Schyns, M. Mitigating the risk of extreme water scarcity and dependency: The case of Jordan. *Water* **2015**, *7*, 5705–5730. [CrossRef]

47. SIWI: World Water Week. *Water and Waste: Reduce and Reuse*; SIWI: Stockholm, Sweden, 2017.

48. Pintilie, L.; Torres, C.M.; Teodosiu, C.; Castells, F. Urban wastewater reclamation for industrial reuse: An LCA case study. *J. Clean. Prod.* **2016**, *139*, 1–14. [CrossRef]

49. Lyu, S.; Chen, W.; Zhang, W.; Fan, Y.; Jiao, W. Wastewater reclamation and reuse in China: Opportunities and challenges. *J. Environ. Sci.* **2016**, *39*, 86–96. [CrossRef]

50. Mo, W.; Zhang, Q. Energy-nutrients-water-nexus: Integrated resource recovery in municipal wastewater treatment plants. *J. Environ. Manag.* **2013**, *127*, 255–267. [CrossRef]

51. Husnain, T.; Liu, Y.; Riffat, R.; Mi, B. Integration of forward osmosis and membrane distillation for sustainable wastewater reuse. *Sep. Purif. Technol.* **2015**, *156*, 424–431. [CrossRef]

52. Geissen, V.; Mol, H.; Klumpp, E.; Umlauf, G.; Nadal, M.; van der Ploeg, M.; van de Zee, S.E.A.T.M.; Ritsema, C.J. Emerging pollutants in the environment: A challenge for water resource management. *Int. Soil Water Conserv. Res.* **2015**, *3*, 57–65. [CrossRef]

53. USAID. *National Strategic Wastewater Master Plan for Jordan*; USAID-ISSP: Washington, DC, USA, 2013.

54. Hussein, H. Tomatoes, tribes, bananas, and businessmen: An analysis of the shadow state and of the politics of water in Jordan. *Environ. Sci. Policy* **2018**, *84*, 170–176. [CrossRef]

55. Water Authority of Jordan. *Groundwater Basins Directorate Open Files*; Water Authority of Jordan: Amman, Jordan, 2016.

56. Tong, L.; Liu, X.; Liu, X.; Zengwei, Y.; Zhang, Q. Life cycle assessment of water reuse systems in an industrial park. *J. Environ. Manag.* **2013**, *129*, 471–478. [CrossRef]

57. Chang, D.; Ma, Z. Wastewater reclamation and reuse in Beijing: Influence factors and policy implications. *Desalination* **2012**, *297*, 72–78. [CrossRef]

58. Yang, H.; Abbaspour, H.C. Analysis of wastewater reuse potential in Beijing. *Desalination* **2007**, *212*, 238–250. [CrossRef]

59. Smith, K.; Liu, S.; Hu, H.Y.; Dong, X.; Wen, X. Water and energy recovery: The future of wastewater in China. *Sci. Total Environ.* **2018**, *637*, 1466–1470. [CrossRef]

60. Li, Y.; Xiong, W.; Zhang, W.; Wang, C.; Wang, P. Life cycle assessment of water supply alternatives in water-receiving areas of the South-to-NorthWater Diversion Project in China. *Water Res.* **2016**, *89*, 9–19. [CrossRef]

61. Anne DE PAZZIS. As Samra Wastewater Treatment Plant A Major Asset for Sustainability in Jordan. Presented at the Stockholm World Water Week, Stockholm, Sweden, 31 August 2017.

62. SAMRA WASTEWATER TREATMENT PLANT JORDAN. Available online: http://www.fogevaporator.com/pdf/samra-wastewater-treatment-plant.pdf (accessed on 18 November 2019).

63. McCarty, P.L.; Bae, J.; Kim, J. Domestic wastewater treatment as a net energy producer–can this be achieved? *Environ. Sci. Technol.* **2011**, *45*, 7100–7106. [CrossRef]

64. Saidan, M.; Khasawneh, H.J.; Aboelnga, H.; Meric, S.; Kalavrouziotis, I.; Hayek, B.O.; Al-Momany, S.; Al Malla, M.; Porro, J.C. Baseline carbon emission assessment in water utilities in Jordan using ECAM tool. Baseline carbon emission assessment in water utilities in Jordan using ECAM tool. *J. Water Supply Res. Technol.* **2019**, *68*, 460–473. [CrossRef]

65. Mehariya, S.; Patel, A.K.; Obulisamy, P.K.; Punniyakotti, E.; Wong, J.W.C. Co-digestion of food waste and sewage sludge for methane production: Current status and perspective. *Bioresour. Technol.* **2018**, *265*, 519–531. [CrossRef] [PubMed]

66. Agyeman, F.O.; Tao, W. Anaerobic co-digestion of food waste and dairy manure: Effects of food waste particle size and organic loading rate. *J. Environ. Manag.* **2014**, *133*, 268–274. [CrossRef]

67. Appels, L.; Lauwers, J.; Degrève, J.; Helsen, L.; Lievens, B.; Willems, K.; Van Impe, J.; Dewil, R. Anaerobic digestion in global bio-energy production: Potential and research challenges. *Renew. Sustain. Energy Rev.* **2011**, *15*, 4295–4301. [CrossRef]

68. Awasthi, M.K.; Pandey, A.K.; Bundela, P.S.; Wong, J.W.C.; Li, R.; Zhang, Z. Co-composting of gelatin industry sludge combined with organic fraction of municipal solid waste and poultry waste employing zeolite mixed with enriched nitrifying bacterial consortium. *Bioresour. Technol.* **2016**, *213*, 181–189. [CrossRef] [PubMed]

69. Chiu, S.L.H.; Lo, I.M.C. Reviewing the anaerobic digestion and co-digestion process of food waste from the perspectives on biogas production performance and environmental impacts. *Environ. Sci. Pollut. Res.* **2016**, *23*, 24435–24450. [CrossRef]

70. Al-Addous, M.; Saidan, M.N.; Bdour, M.; Alnaief, M. Evaluation of Biogas Production from the Co-Digestion of Municipal Food Waste and Wastewater Sludge at Refugee Camps Using an Automated Methane Potential Test System. *Energies* **2019**, *12*, 32. [CrossRef]

71. Dai, X.; Li, X.; Zhang, D.; Chen, Y.; Dai, L. Simultaneous enhancement of methane production and methane content in biogas from waste activated sludge and perennial ryegrass anaerobic co-digestion: The effects of pH and C/N ratio. *Bioresour. Technol.* **2016**, *216*, 323–330. [CrossRef]

72. Wan, J.; Gu, J.; Zhao, Q.; Liu, Y. COD capture: A feasible option towards energy self sufficient domestic wastewater treatment. *Sci. Rep.* **2016**, *6*, 25054. [CrossRef]

73. Gu, Y.; Li, Y.; Li, X.; Luo, P.; Wang, H.; Robinson, Z.P.; Wang, X.; Wu, J.; Li, F. The feasibility and challenges of energy self-sufficient wastewater treatment plants. *Appl. Energy* **2017**, *204*, 1463–1475. [CrossRef]

74. Hagos, K.; Zong, J.; Li, D.; Liu, C.; Lu, X. Anaerobic co-digestion process for biogas production: Progress, challenges and perspectives. *Renew. Sustain. Energy Rev.* **2017**, *76*, 1485–1496. [CrossRef]

75. Ronteltap, M.; Maurer, M.; Gujer, W. The behaviour of pharmaceuticals and heavy metals during struvite precipitation in urine. *Water Res.* **2007**, *41*, 1859–1868. [CrossRef] [PubMed]

76. Uysal, A.; Yilmazel, Y.D.; Demirer, G. The determination of fertilizer quality of the formed struvite from effluent of a sewage sludge anaerobic digester. *J. Hazard. Mater.* **2010**, *181*, 248–254. [CrossRef]

77. Antakyali, D.; Kuch, B.; Preyl, V.; Steinmetz, H. Effect of Micropollutants in Wastewater on Recovered Struvite, Proc. *Water Environ. Fed. Nutr. Recovery Manag.* **2011**, *1*, 575–582. [CrossRef]

78. Jabr, G.; Saidan, M.; Al-Hmoud, N. Phosphorus Recovery by Struvite Formation from Al Samra Municipal Wastewater Treatment Plant in Jordan. *Desalin. Water Treat.* **2019**. [CrossRef]

79. EPA. *Guidelines for Water Reuse, U.S.*; Environmental Protection Agency: Washington, DC, USA, 2004.

80. Saidan, M.N.; Al-Yazjeen, H.; Abdalla, A.; Khasawneh, H.J.; Al-Naimat, H.; Al Alami, N.; Adawy, M.; Jaber, M.S.; Sowan, N. Assessment of on-site treatment process of institutional building's wastewater. *Processes* **2018**, *6*, 26. [CrossRef]

81. Bixio, D.; Wintgens, T. *Water Reuse System Management—Manual AQUAREC. Directorate-General for Research*; European Commission: Brussels, Belgium, 2006.

82. Huber, M.M.; Gobel, A.; Joss, A.; Hermann, N.; Loffler, D.; McArdell, C.S.; Ried, A.; Siegrist, H.; Ternes, T.A.; von Gunten, U. Oxidation of pharmaceuticals during ozonation of municipal wastewater effluents: A pilot study. *Environ. Sci. Technol.* **2005**, *39*, 4290–4299. [CrossRef] [PubMed]

83. Feigin, A.; Ravina, I.; Shalhevet, J. *Irrigation with Treated Sewage Effluent*, 1st ed.; Springer: Berlin/Heidelberg, Germany, 1991.

84. Brissaud, F. Low technology systems for wastewater treatment: Perspectives. *Water Sci. Technol.* **2007**, *55*, 1–9. [CrossRef] [PubMed]

85. Ghermandi, A.; Bixio, D.; Traverso, P.; Cersosimo, I.; Thoeye, C. The removal of pathogens in surface-flow constructed wetlands and its implications for water reuse. *Water Sci. Technol.* **2007**, *56*, 207–216. [CrossRef] [PubMed]

86. Oron, G.; Gillerman, L.; Buriakovsky, N.; Bick, A.; Gargir, M.; Dolan, Y.; Manor, Y.; Katz, L.; Hagin, J. Membrane technology for advanced wastewater reclamation for sustainable agriculture production. *Desalination* **2008**, *218*, 170–180. [CrossRef]

87. Hyun, K.S.; Lee, S.J. Biofilm/membrane filtration for reclamation and reuse of rural wastewaters. *Water Sci. Technol.* **2009**, *59*, 2145–2152. [CrossRef]

88. Xu, P.; Drewes, J.E.; Bellona, C.; Amy, G.; Kim, T.U.; Adam, M.; Heberer, T. Rejection of emerging organic micropollutants in nanofiltration-reverse osmosis membrane applications. *Water Environ. Res.* **2005**, *77*, 40–48. [CrossRef] [PubMed]

89. Norton-Brandao, D.; Scherrenberg, S.M.; van Lier, J.B. Reclamation of used urban waters for irrigation purposes—A review of treatment technologies. *J. Environ. Manag.* **2013**, *122*, 85–98. [CrossRef] [PubMed]

90. Cano, A.; Cañizares, P.; Barrera, C.; Sáez, C.; Rodrigo, M.A. Use of low current densities in electrolyses with conductive-diamond electrochemical e oxidation to disinfect treated wastewaters for reuse. *Electrochem. Commun.* **2011**, *13*, 1268–1270. [CrossRef]

91. Wei, X.Z.; Gu, P.; Zhang, G.H. Research progress in the new model of pollutants in water and their removing methods. *Ind. Water Treat.* **2014**, *34*, 8–12.

92. Bernabeu, A.; Vercher, R.F.; Santos-Juanes, L.; Simon, P.J.; Lardin, C.; Martinez, M.A.; Vicente, J.A.; Gonzalez, R.; llosa, C.; Arques, A.; et al. Solar photocatalysis as a tertiary treatment to remove emerging pollutants from wastewater treatment plant effluents. *Catal. Today* **2011**, *161*, 235–240. [CrossRef]

93. Pedrero, F.; Albuquerque, A.; Amado, L.; Marecos do Monte, H.; Alarcón, J. Analysis of the reclamation treatment capability of a constructed wetland for reuse. *Water Pract. Technol.* **2011**, *6*, wpt2011050. [CrossRef]

94. Cirelli, G.L.; Consoli, S.; Di Grande, V. Long-term storage of reclaimed water: The case studies in Sicily (Italy). *Desalination* **2008**, *218*, 62–73. [CrossRef]

95. Yi, L.; Jiao, W.; Chen, X.; Chen, W. An overview of reclaimed water reuse in China. *J. Environ. Sci.* **2011**, *23*, 1585–1593. [CrossRef]

96. Waterwise Queensland. Manual for Recycled Water Agreements in Queensland. The State of Queensland. Environmental Protection Agency. 2005. Available online: http://www.epa.qld.gov.au/waterrecyclingagreement (accessed on 12 November 2019).

97. COAG. *Communique August*; Council of Australian Governments: Canberra, Australia, 2003.

98. Di Carlo, A.; Sherman, A. The Reclaimed Water Agreement Manual, Russell Kennedy Solicitors, Melbourne. 2004. Available online: http://www.clw.csiro.au/priorities/urban/awcrrp/stage1files/AWCRRP_8%20Final_27Apr2004.pdf (accessed on 12 November 2019).

99. Moore, W.; Gould, J. Shoalhaven Reclaimed Water Management Scheme—How a small region built a large water recycling scheme. In Proceedings of the Water Recycling Australia, 2nd National Conference, Brisbane, Australia, 1–2 September 2003.

MDPI
St. Alban-Anlage 66
4052 Basel
Switzerland
Tel. +41 61 683 77 34
Fax +41 61 302 89 18
www.mdpi.com

Water Editorial Office
E-mail: water@mdpi.com
www.mdpi.com/journal/water